The LOVED DOG

The LOVED DOG

THE GENTLE WAY TO TEACH YOUR DOG GOOD MANNERS

TAMAR GELLER

WITH ANDREA CAGAN

3 5 7 9 10 8 6 4

Published in 2007 by Vermilion, an imprint of Ebury Publishing
A Random House Group company

First published in the USA in 2007 by Simon Spotlight Entertainment

The Random House Group Limited Reg. No. 954009

Addresses for companies within the Random House Group can be found at www.randomhouse.co.uk

A CIP catalogue record for this book is available from the British Library

The Random House Group Limited supports The Forest Stewardship Council (FSC), the leading international forest certification organisation. All our titles that are printed on Greenpeace approved FSC certified paper carry the FSC logo.
Our paper procurement policy can be found at
www.rbooks.co.uk/environment

Designed by Margaret Gallagher

Printed and bound in Great Britain by
CPI Mackays, Chatham, ME5 8TD

ISBN 9780091922252

Copies are available at special rates for bulk orders. Contact the sales development team on 020 7840 8487 for more information.

To buy books by your favourite authors and register for offers, visit www.rbooks.co.uk

Love the animals, love the plants, love everything.

If you love everything, you will perceive the divine mystery in things.

– Fyodor Dostoyevsky from The Brothers Karamazov

Acknowledgments

Gratitude is key in celebrating life and there are many people (and dogs) who have helped make this book happen.

I'd like to thank Carrie Cook, for convincing me to write a book; Andrea Cagan, for being so open and loving to me and to animals; and Jen Bergstrom and Ursula Cary at Simon Spotlight Entertainment, for feeling so strongly about dogs.

To Eric, Tania, Lily, and Wee – thank you for the love, inspiration, and 'can do' attitude – and Wee, for your amazing cooking when I came up for air!

To Lilach Barka and my team at The Loved Dog – thank you for helping with everything that I asked for, and particularly for showering my doggies with love while I was spending those long hours by the computer.

To my dear Richard Beckerman, who stayed up with me until 5 AM while I was writing – thank you for your understanding, support, and love.

To Wayne Pacelle and Kathy Bauch – thank you for your unparalleled dedication and support of me and my work.

To my Dad – thank you for sharing your love of animals, and for telling me to always follow my passion in life and never compromise.

To my beloved Clydie and Dukey – thank you for the constant love and for making me go on a hike every day!

And finally, a big thanks to all the dogs and their families who invited me in and shared their stories with me. You are my teachers.

Contents

Foreword		xi
Introduction		xiii
Part One	Stranger in a Strange Land	1
Chapter One	A Living Dream	3
Chapter Two	The Loved Dog	11
Chapter Three	My Two 'Sons'	17
Chapter Four	Pain and Pleasure	23
Chapter Five	The Seven Basic Needs	31
Part Two	Play-Training	45
Chapter Six	A Common Language	47
Chapter Seven	Using Treats	61
Chapter Eight	Socialization	71

Chapter Nine	The Power of Sit	87
Chapter Ten	Jumping	99
Chapter Eleven	The Best Wolf Game	107
Chapter Twelve	The Power of Toys	117
Chapter Thirteen	Come	125
Chapter Fourteen	Backing Off	131
Chapter Fifteen	Chilling Down	137
Chapter Sixteen	The Secret of Stay	145
Chapter Seventeen	Housebreaking	153
Chapter Eighteen	Walking	165
Chapter Nineteen	Barking	173
Chapter Twenty	Celebrate!	179

FOREWORD

When Tamar Geller asked me to write the foreword for this book, she did not need to reward me with a treat or some verbal affirmation as an inducement. I've come to know Tamar as an inspired teacher and life coach for dogs and their people, and I have seen first-hand the extraordinary results of her work. And more than that, the beliefs that underlie her work, and her life, are perfectly aligned with the mission of The Humane Society of the United States – and it's a pleasure, on personal and professional levels, to be associated with this work.

Respect for all beings, kindness not violence, an understanding of the intrinsic value of animals, appreciating a dog's dog-ness, building bonds between people and dogs that will last a lifetime – that's what keeps dogs with their families, and out of our nation's shelters for homeless animals, for a lifetime.

That's also what makes this book so much more than a dog-training manual. To be sure, you'll find a trove of practical information and tips inside. But Tamar won't tell you how to teach your dog to sit so much as teach *you* how to inspire your dog to *want* to sit. Like the best coaches, she'll help bring you and your dog together as a successful team.

At a time when people rightly recognize dogs as members of their families, dogs should indeed be treated with the respect and kindness we'd give any other member of the family. There is wisdom in Tamar Geller's words and approach.

The *how* of her method is simple, and the subtitle of the book says it all: The Gentle Way to Teach Your Dog Good Manners. The *why* of her method is even simpler: all dogs deserve to be loved dogs. And that is a worthy goal indeed.

– Wayne Pacelle, president and CEO, The Humane Society of the United States

INTRODUCTION

How the heck do you flush this thing? I asked myself. I was in a makeshift bathroom in a small guest house in Bangkok, standing over a toilet with no seat, when I realized that there was no conceivable way to flush.

I walked into the living room, my face blushing with embarrassment. 'Can you help me?' I shyly asked the elderly woman I'd met for the first time just a moment before. 'I couldn't flush the toilet. I'm not sure what to do.'

The lady of the house cracked a yellow-toothed smile and asked me in heavily accented English, 'You see bathtub and bucket next to toilet?'

I nodded.

'Fill up bucket with water and throw in toilet. That make it flush. Then you use water to wash.'

I thanked her and returned to the bathroom, where I practised my unusual and new-found knowledge. Still embarrassed, I shuddered to think what would have happened if I'd been caught by the person to follow me, or if this woman had not understood me.

Wait just a minute, you're probably thinking. *What do your travel adventures have to do with dog training?*

The answer is: everything! When we bring a new dog into our home, he or she is like an alien entering a foreign land, with no idea how to fit in, how things work, and – worst of all – no way to ask for help or instructions. Then, to add insult and pain to confusion, when they do their business in the wrong place (according to us), they get their highly sensitive noses shoved into their own excrement, smacked with a newspaper, or thrown outside in shame, leaving them to wonder if these new people are safe to be around and if the natural act of going to the bathroom is wrong. Thank goodness the woman in Bangkok didn't do that to me!

Imagine if someone only spoke gibberish to you, yanked your neck, and smacked or caged you when you failed to understand. Would you feel love toward her? What's more, the subsequent anxiety, fear, and anger you felt would make it even more difficult to calm down or to understand what you were supposed to do.

I have observed the horrendous methods (in my opinion) that many trainers use to train dogs. These trainers claim punishment is necessary and effective, but believe me, violence is never appropriate – even when it is disguised under some very creative-sounding terms! Each time a trainer wields punishment against a dog, a disturbed animal is created. If you blindly yell and yank on your dog's leash, you're acting like an out-of-control, angry lunatic – not someone trustworthy – and the dog will end up being afraid of you.

I don't believe in ever using stress, intimidation, violence, alpha rollovers, or fear. This means no prong collars, choke chains, physical correction, or anything else I call 'Spanish Inquisition methods', which are pain-, stress-, or fear-based. On the contrary I have found that games are the best teaching tools. The foundation of all dog coaching is as simple as changing what your dog associates with pain and what he associates with pleasure. Imagine indicating to your dog that the training session is over, and instead of his exhaling a sigh of relief and running to hide under the bed, he sits in front of you politely, his tail wagging, asking for more!

To me, teaching a dog 'to obey' sounds militant, as if the dog has no choices and he *will* do what you command of him – or else! Successfully training your dog is not about getting him to obey with no questions asked. Teaching manners to your dog and building a strong bond will bring you the results you want. Your dog will *want* to come to you. After all, you are his benevolent leader, and he's proud to be your best friend!

Each time you teach a dog through 'play-training', set achievable short-term goals and celebrate his successes. Each time you replace a no with a yes, you are building an intense and unconditional friendship that will bring both of you endless satisfaction and joy. I can't tell you how many times clients look at me with amazement and say, 'This is so easy and it makes so much sense. Why didn't I think of that?' Or 'I didn't know my dog was so smart. After all, she failed obedience school!'

The difference between teaching obedience through force and teaching manners through kindness is vast. The bottom line is that dogs fulfill our basic needs for connection, love, and acceptance. They are happy to give us these things, even on our worst days. We can show our gratitude and appreciation by consistently meeting their own seven basic needs, which are detailed in chapter 5. Let me impress upon you that if you are looking for an abundance of loyalty and devotion, your dog is eager to give these things to you. No matter the state of your life or how you look at any given time, your dog will be there, reminding you how great you are and encouraging you to smile.

This book is about the ins and outs of dog training, which are play-training, coaching, and teaching manners. After all, if a dog is going to become a part of your family, you need to teach her as you would teach your child to be polite. At the same time, however, you need to satisfy your dog's untamed 'wolf nature'.

In the pages to come, I'll share with you how to use my non-violent methods. I'll also show you how easily and quickly you can teach your dog to sit, lie down, quit jumping on guests, and stop begging at the table. I'll share the secret of how to develop a common language so you can communicate effectively with your dog. We'll also discuss how diet and exercise impact on your dog's attitude and ability to learn good manners and positive behaviors.

At the same time, this book is about more than just teaching your dog to sit, stay, and roll over. It's about life, how we choose to live it, and how much we can learn from our dogs. Success in dog coaching comes from consistency and your commitment to do what it takes to create a beautiful, productive, safe, and loving relationship. Then you can unleash the magic in him for all the world to see.

If you follow my play-training methods daily, you and your dog will see how easy and pleasurable dog coaching can be. The amazing results will thrill you both! I want you and your dog to smile at each other every single day and ask yourselves, 'How did we get so lucky?'

Some of my methods may seem completely different from what

you may have been told about dog training. They are! They are also faster to teach and more fun. I ask you to keep an open mind. Your dog will love you for it.

My love for dogs and my desire to communicate with them began in my childhood. However, it wasn't until 1987 that I realized training dogs would become my life's purpose. I spent several months in the Israeli desert, at a field school and research facility. There I began observing wolves in the wild.

What I saw fascinated and amazed me. I witnessed the older wolves integrating the younger ones into the wolf culture, correcting behavior through game playing, body postures, facial expressions, tug-of-war, hide-and-seek, baring teeth (which I call 'visiting the dentist'), threatening growls, and cold shoulder treatments. There was no violence whatsoever! Although it all appeared quite fierce and dangerous, never once did I see one wolf damage another in an effort to socialize a pup into the pack or to get back at someone for a grievance.

During this time a vivid, recurring dream began waking me in the night, telling me that training dogs, using the wolves' natural methods, was my future. Thus began my lifelong career as a non-violent life coach for dogs and their owners. In other words, I have dedicated my life – personally, professionally, and spiritually – to being part of a natural and harmonious world. I believe that peace begins with interactions among our close circles and families, including our pets. As author Henry Salt wrote in his book *The Creed of Kinship*, 'The basis for any real morality must be the sense of kinship between all living things.'

Among the millions of dogs and puppies that are dumped into

shelters, and the estimated three million that are put to sleep every year, about sixty percent are under one year old. How can a nine-month-old puppy, who has been ignored, poorly trained, or violently disciplined, *not* misbehave or show aggression? Why would he *not* defend himself with whatever resources he has?

I need only to look into the big, wet eyes of a dog to understand that violence is not what they are about. Instead I see the awe, mystery, joy and hope that I want to nurture and appreciate. Dogs' natures are both wild and adventurous, with great love at their core, and the more you teach them in a loving, playful way, the more they will want to learn.

Although dogs are related to wolves, their dual natures suggest that they also behave like toddlers, which is the aspect that creates the magical quality of dogs. We need to help them achieve a balance between the wolf and toddler aspects of their personalities, without breaking their wild spirits. We must remember that each dog has an individual character with likes and dislikes. Whether they are purebred or the most creative combination of mutts, they experience frustration and joy at different levels. In essence, all dogs are sensitive, emotional, responsive animals that love a good leader and are hungry to learn and have fun. Their natural optimism is contagious, and their loyalty is unwavering. For these reasons, they are a blessing – magical gifts from God and extraordinary resources for healing and companionship. However, we must learn how to bring out these magnificent qualities.

My goal is to speak up and help disenfranchised beings in this world who have no voices to speak for themselves. I believe we all have a choice about whether we live our lives through love or fear.

My play-training is based on love because that is how I have chosen to live life.

I am not under the illusion that my beliefs are the only ones to believe in or that my way is one hundred percent true for everyone. But I ask you to keep an open mind, an open heart, and take from my experiences whatever makes sense to you. For me, if it's not about love, I'm not interested. I hope you feel the same way.

Part One

Stranger in a Strange Land

Chapter One

A Living Dream

A massive dust cloud rose upward into the night sky, completely obscuring the towering Edom mountain range in the far distance. I had visited this location many times before, and I still marveled at the sheer mountain bluffs rising straight up out of the flat, sprawling expanse of wheat-colored desert. I was keenly aware of night noises as the distinct sounds of crickets, a variety of exotic bugs, bird calls, and mysterious cracklings in the brush filled the air with a disharmonious melody.

I loved the desert and was comfortable with the natural noises. But why in the world were the wolves creating this disturbance that was sending sand and dust clouds up into the atmosphere? By now, I'd watched this wolf pack many times, but I had never seen this kind of activity before. Hardly breathing, I waited for the dust to disperse, scared to death of what I might see when the air cleared.

Crouching, I retreated farther back into the brush and glanced at my bicycle, my sole means of transportation out of here. Sleeping quarters were a good half-hour ride away – provided I didn't get a flat tire from the craggy desert floor and have to patch it in the dark. I continually scanned the earth around me, always on the lookout for the poisonous snakes and deadly yellow scorpions, which were indigenous to the area.

I slowed my breath, reminding myself that I was safe. A man called Zev ('Wolf' in Hebrew, an apt nickname for a man involved in wolf research) had told me that as long as I stayed a good distance away from the wolves I was observing, they would leave me alone. That was no problem. Approaching a pack of wild wolves was the last thing I wanted to do. Still, the thudding of their padded feet on the packed desert earth made me afraid. I calmed myself down, remembering Zev's assurance that wolves were not interested in humans. He said that if they saw me coming too close, they would most likely leave the vicinity before I did. But the dust storm I was watching concerned me, since I didn't understand what was going on.

As the dust diffused, I caught sight of the back legs of a small brown wolf, tearing across the desert floor. His entire body came into view, with the agility and tautness in his legs visible. He suddenly veered off, causing the dust to fly in the opposite direction, exposing several other wolves, slightly smaller than he was, who were chasing him. These were not the huge gray wolves that you see in the movies. Asian wolves, like the ones in front of me, are smaller and brown, about the size of a coyote, with less hair than a gray wolf. Still, they are capable of inflicting a lot of damage, and I feared they were ganging up on the alpha wolf, who was distinguishable by certain facial markings.

Was I observing a change in regime? Would it be bloody? I stared out at the vast desert surrounding me, feeling very small. Untouched by man for the most part, and nearly void of vehicles, the sprawling desert acreage was home to the Chatsevah Field School, a wilderness research facility in the Arava Desert. From the moment I first arrived at the field school, I'd fallen in love with this pristine land – despite the punishing heat, the stinging insects, and the remote location. This was nature at its purest, where animals were observed but never hunted by a group of people who profoundly respected the creatures who lived there.

This field school was not a resort where you could make a reservation, check in, and order room service. It was a facility for animal, insect, reptile, and bird researchers. I needed a job in order to be there, so my friend introduced me to a professor who was researching Israeli desert birds called *zanvanim*, or 'Palestine babblers'. He needed some help and it was swiftly arranged that while the professor was busy teaching in Tel Aviv, I would stay at Chatsevah as his research assistant, observing the *zanvanim* twice a day, every day, taking notes on their behavior.

Each morning I woke up before sunrise and rode my bicycle to a particular tree where the birds nested. As much as I hated the sound of the alarm clock in the dark, I absolutely loved being out before the world woke up.

After taking notes on the *zanvanims'* behavior, when it was too hot to do anything else, I would ride back to catch some sleep. The rest of the day was unstructured until an hour or so before sunset, and back I went to the tree where the birds lived. Each time I watched the sun rise or set against the golden mountains in the distance, I felt as if I was on a movie set. The truth was, with the exception of a few hours a day when I watched the birds, I was lonely and bored.

One evening I saw a Jeep pull up outside the facility with a dead ibex (an Israeli antelope) in the back. Flies buzzed around the smelly carcass. I met Zev, the driver of the Jeep, for the first time. 'What's going on?' I asked him.

'I research wolves,' Zev told me. 'We found this dead ibex, and we're heading to a feeding station to drop it off.'

'Can I come?' I asked eagerly.

'Sure,' he said. 'Hop in.'

I sat in the back of the Jeep, my feet awkwardly resting on the seat in front of me to avoid touching the ibex. It was about 9 PM when we pulled up to an area designated as a wolf feeding station. We dragged the carcass out of the Jeep and dropped it heavily onto the earth. Then we drove a safe distance away, but where we could still see the ibex, and we waited. We had flashlights, but we didn't need them because the night was lit up by millions of brilliant stars.

We waited quietly for several hours, seeing nothing. Then I caught sight of the dark silhouette of a slinking animal. Another wolf followed close behind, and the two sniffed the ground, heading for the dead ibex. More arrived, about eight in all. I knew they smelled us there, but they were not focused on us. They were much more interested in the bloody, dead gift we had left for them. While the pack salivated and waited, one wolf (I would learn later that he was the alpha) was the first to bury his face into the carcass's stomach, tearing through the soft flesh. He pulled out the lining of the stomach, blood dripping from his snout, and ate it. The others followed, ripping the flesh apart, exposing the organs. They would tear out a chunk of meat, run a short distance away, devour it ravenously, and then go back for more. The stench of the innards was putrid, something I will never forget. But to them it appeared to be the finest delicacy on Earth.

Later Zev showed me several other feeding stations where they left dead animals they occasionally found in the desert. I began returning to those places on my own, observing the wolves any night I could find them. What had started as a way to relieve my boredom had turned into the best part of my day. Each night I crouched in the bushes, waited, and watched. As I observed these extraordinary, feral creatures living their lives naturally, I was completely engrossed. I took it all in and felt the thrill of the magic that was occurring right in front of my eyes.

On the night of the dust cloud, I held my breath. The wolves ran like a pack of lunatics and closed in on the alpha wolf. I had no idea what to expect. As I fought the urge to cover my eyes, the chased wolf stopped abruptly – as if someone had suddenly yanked a pair of invisible reins. When he stopped, so did the others. They stared into his eyes, breathless and panting. I could have sworn that then the alpha wolf smiled. I saw his teeth and a flash of light in his eyes. He bent low on his back haunches, his front legs extended long in front of him in a 'play bow'. Then he took off, chasing the wolf closest to him. He had turned the tables, and predator and prey flew across the sand – this time with the roles reversed.

The alpha wolf stopped after a few minutes, seemingly tired of the game. He sniffed the ground and lay down. He looked bored as the wolf he had been chasing crept slowly toward him, his tail wagging low, his head down, his ears back. He started to clean the leader's ears, licking and grooming. The alpha wolf allowed it, as if he were accepting an apology. Order was established once again. The chase that had looked like a fight to the death was over. It had only been a game, albeit a fierce-looking one. I later came to realize that this behavior not only taught the wolves to hunt their prey, but it also helped to establish who was boss, a crucial aspect of their society. No one had been truly threatened, and no one got hurt.

As the wolf pack rested and groomed one another after their immense energy output, I heaved a sigh of relief. I suddenly understood something. The alpha wolf, the one who was originally being chased, had never been the victim. Not for an instant. He had the upper hand the entire time. He taunted the pack with a challenge and then took off, racing around in random patterns, as if to say, 'Catch me if you can.' They couldn't. He was faster, smarter, and more agile than the rest – the reasons he was the alpha wolf. If anyone in the pack was not sure, he had proven it, in no uncertain terms.

I learned many other things from the wolves and their game playing, such as when they were engaged in tug-of-war. A wild and dangerous-looking activity, it was one of their favorite methods for establishing hierarchical positions. Most of the challenges took place among wolves who were vying for the second and third positions in the pack, not the top spot that was already established. A wolf would grab one end of a leftover animal leg or a tree branch between his teeth, and another wolf would bite onto the other end. They salivated while they pulled on their respective sides, with sometimes as many as three or four wolves joining the play battle. Whoever tugged harder than the rest ended up not only with the branch, but with the respect of the others, without having to use any violence.

That night I had the first in a series of recurring dreams that would ultimately shape my future. In the dream I heard a calm, authoritative voice tell me: *You must work with dogs.*

I awoke and sat upright, staring at the white walls around me. The words in my dream were clear, and I was eager to comply. But what did they really mean? I felt that I had been touched by something otherworldly, and I knew that this dream held deep significance.

But what action was I supposed to take?

When my time at Chatsevah was over, I threw a large backpack over my shoulders and traveled to southeast Asia for close to a year, temporarily postponing the dream. I would recall it in the future, though, as I started my own business as a life coach for dogs and their owners. I would base my training methods on the way the wolves interacted with one another, integrated their cubs into the pack, and respected their leader beyond anything or anyone else.

Chapter Two

The Loved Dog

In 1996, I founded The Loved Dog Center, a cage-free doggy day care and boarding facility in West Los Angeles. The idea of doggy day care first came to me in Brentwood, California, where I was called to a very fancy home to help a couple with their supposedly 'unmanageable' dog.

'When we get home,' the owners had told me, 'he drives us insane. He jumps and whines and nips, and he gets so crazy, there's nothing we can do with him.'

I arrived at their home one evening at nine. This was when they usually returned from work to find Mikey, their Labrador mix, in a manic frenzy at the front door. 'What has he done today?' I asked them.

'Well,' they said, 'we're not really sure. He had the run of the house and our huge backyard and plenty of food, but he still makes us insane at night. We're so exhausted. Can you train him to stay put?'

I stared at them. 'Are you telling me that you leave this dog alone from eight in the morning until nine at night? He has no companionship, he gets no exercise, and you expect him to sit quietly when you get home? You want me to teach him to sit, stay, and lie down right now? Dogs are social animals, and it isn't fair to the dog to even try to coach him. How about calling a neighbor dog for a playdate with him? Or hiring a dog walker?'

They looked at each other blankly and then back at me. 'Can't *you* just take him?' they asked.

They did not understand that for a dog, being social is not a luxury but a basic need. Think about it – even for a hardened criminal, solitary confinement is one of the worst possible punishments. This couple had, for all intents and purposes, condemned their innocent dog to a lifetime sentence in solitary confinement. They did not understand that after so much isolation, boredom, and loneliness (extremely stressful for any pack animal), finally having some company was a great relief. It makes perfect sense that the eventual result was a wild and needy pet, by no means the fault of the dog.

That evening, as I pondered Mikey's fate, I envisioned a safe place for dogs, with no cages or bars, where people could leave their dogs while they were at work or on vacation. After careful screening for aggression, the dogs would get to be part of a pack. They could let out their wolf natures and use up their vast storehouses of energy by playing games with one another under full supervision and with love. When they were picked up at the end of the day, tuckered out and ready to sleep, everyone would be happy.

When I first told a friend my idea, she laughed. 'You have to be kidding. Day care for dogs? That's the nuttiest thing I've ever heard.'

In my heart, though, I knew I was responding to a need – not some over-the-top spoiled behavior that you would only see in Hollywood. Doggy day care and overnight stays in a completely cage-free environment had never been done before (at least not in Los Angeles), but that was part of the attraction. I've always loved a good challenge. Even my 'smart' friends told me that my idea was pathetic, but all that did was add fuel to the fire.

I leased a two-storey, six-thousand-square-foot warehouse and faced the bureaucracy – the animal regulation board and the city zoning commission – to sell them on the idea of a totally cage-free kennel. It was an ordeal to convince this very strict board to accept such a new concept, but in the end, I got my permit. I had the first legal cage-free dog kennel in Los Angeles!

When I first opened the doors, I had nothing but a huge space and two-and-a-half-day dogs – two that stayed all day and one that stayed for half a day. *What have I done now?* I thought. I did it all back then – answering phones, cleaning the warehouse, and supervising the dogs as they played on the slides, or lay around on the soft couches, sleeping and dreaming with their paws twitching. Sometimes I even drove them home at the end of the day. But no matter how tired I was, I was thrilled to see the direction in which my life was going.

After numerous meetings with skeptical veterinarians and enthusiastic dog lovers, I had a bunch of happy customers, both two- and four-legged. Kiku, a Basenji who has been with us since our first day, started tapping his owner's shoulder with his front paw whenever they got to a certain point in the road. He knew he was heading to The Loved Dog, and he made sure his owner didn't forget to turn. Ginger Bosley, actor Tom Bosley's shepherd mix, counted the minutes by the front door, pressing her nose against the glass until she saw my SUV turning into her carport to pick her up.

One of our dogs, Jack, a Jack Russell terrier, waited for his owner to pull up in front of The Loved Dog. Then he'd jump out of the car window and rush to the front door, crying to be let in. A school teacher said that her dog was so happy and exhausted when she picked him up at the end of the day, he conked out on the backseat immediately. And a lawyer told us that his dog was so tired out, she slept right through dinner. Since everyone's needs had been met, people now spent their evenings together with their dogs in a relaxed and loving way, instead of returning to 'doggy-redecorated' homes, or anxious and depressed pets.

These days, we house about eighty dogs, either for the day or for overnight boarding. I have a fantastic team of dedicated employees, and they show up with great enthusiasm and anticipation. The owners are thrilled to leave their pets to join a pack of happy animals. It's such a far cry from dropping off a fearful, trembling dog, tail between his legs, head bowed in terror, getting dumped in a cage where he loses his appetite while he's counting the minutes till his owner returns.

I installed a large window so owners can look in and watch their dogs playing happily with their own kind. They let out their wolf natures in a safe place where they are watched over by caregivers who act as leaders of the pack. It's a moving experience for an owner to see her dog as part of nature and not just as her 'baby'. I constantly have to replenish the tissue box by the window, because people really get emotional when they see their dogs playing freely and with wild abandon.

One morning, in my SUV on the way to The Loved Dog Center, I happened to hear a popular female talk-show psychologist on the radio. She was taking a call from a woman with a two-year-old daughter, which caught my attention.

'What should I do about my father?' the distressed woman asked. 'He calls my two-year-old daughter to come to him, and when she doesn't obey immediately, he hits her.'

'That can never happen again,' the psychologist said with conviction. 'That is totally unacceptable behavior, and you need to make sure your daughter is never, ever alone with your father again.'

I pulled up to the front door of my building, turned off the radio, and parked. My two dogs, Clydie and Dukey, were wagging their tails so hard, I thought they might knock each other over. I opened the door to let the dogs out onto the sidewalk, and they rushed toward the entrance to my facility, flying up the stairs together just behind me. I was eager to see my team of employees and our four-legged visitors, who always greeted me like a herd of happy buffaloe. But I just couldn't stop thinking about the woman on the radio. Oftentimes, on the advice of dog trainers, books, or friends who claim to know how to train dogs, people bark out commands as if their pets were in the military. The punishments these poor animals receive for not understanding English or not acting quickly enough are often swift and brutal. That's no way to treat any living creature. Even wolves in the wild, feral and in their primal state, do not treat their own kind in such a violent, short-sighted way. I wanted people to react to a dog being hit with the same clarity and outrage the psychologist on the radio had: 'That can never happen again.'

Chapter Three

My Two 'Sons'

Let me take a moment here to introduce you to my boys, Clyde and Duke, whom I love passionately and unconditionally.

Clyde, a Doberman mix, came to me through a veterinarian friend. 'There's a dog here who really needs you, Tamar,' he said over the phone.

My heart sank. Three weeks earlier, I had put down Michaela, my Rhodesian Ridgeback mix, and I was still mourning the loss. 'I think it's too soon,' I told him.

'He's exactly the kind of dog you like,' he persisted. 'He has short hair and floppy ears, and besides, he really needs you.'

I reluctantly drove to the vet with my friend Tom, unable to resist seeing this dog but determined that I would absolutely *not* be taking him home. I was still grieving for Michaela, and I wasn't sure if I was ready to open my heart to another dog. But when the poor guy was brought out, still puppyish at ten months old, he lay down, rolled over on his back, and peed all over the place. He was terrified with a capital *T*. My heart melted, and although I wasn't sure if I could muster the energy and enthusiasm to build his confidence, I took him home anyway. When I put him into the car, his fear escalated to such a high level, he started to drool uncontrollably.

A few weeks earlier, someone had thrown him from a speeding car into the middle of a freeway. It was a miracle he had not been run over. Luckily a dog walker witnessed the crime and took him to the vet. Now, as my friend drove and I sat in the backseat with Clyde, he panicked, and suddenly tried to jump out of the window of the moving car. I grabbed him by the back legs, pulled him in, and for a long time after that, no windows were ever cracked open whenever Clyde was riding in the car. It was up to the air conditioner to provide our only fresh air!

I lived near a lake, so I took Clyde swimming every day. I also took him on daily hikes, and he seemed to really like the outdoors. He was still too fearful for any formal coaching, but when I showed Clyde a tennis ball, he turned out to be really good at playing fetch. Each time he caught the ball in the air and brought it to me, I made a fuss, like he had just won the gold medal at the doggy Olympics. I taught him the name of his girlfriend, Ruby, a neighboring dog with whom he loved to play. I gave him treats, cheered him on, and started to teach him English. I repeated the word 'fetch' when we played ball. And he taught me a few games with the ball that he made up himself. I was becoming the proudest mama in the world! I was building his self-esteem the way you would with a toddler. We were developing a mutual understanding, and I was falling in love with him – and then he got really sick.

One morning when I threw the ball for him, he wasn't interested. Normally that would not be cause for alarm, but Clyde had always been much too obsessed with the tennis ball. I knew something was wrong, so I took him to the vet. After some initial tests, he left with a clean bill of health. I was confused – until I spotted blood in his urine. Another vet visit revealed that Clyde had haemophiliac anaemia, a blood condition from which dogs rarely heal. We ended up in the office of a veterinarian specialist, Dr Feldman, who admitted Clyde to the hospital late on a Friday afternoon.

He put Clyde on an IV and gave him blood transfusions. I was with Clyde the entire weekend and every day after that, watching him waste away to just skin and bones. I cried and prayed, his friends and mine came to see him, and then a miracle began to happen. One day a month later, his blood platelets were high enough for me to take him home. He wasn't well, though. He could barely lift his head. I cooked him organic food, took him to an acupuncturist, and let him watch his doggy friends play together.

A few more weeks of good food and treatments (and twenty thousand dollars later), I had my boy back, healthy and strong. Today, at almost nine years old, he goes with me when I give talks at schools, taking his teacher's assistant role very seriously. Everyone marvels at how human he seems and how youthful and playful he is. In the morning, Clyde still greets me – at attention, with the ball in his mouth, ready to play. He holds no grudges, he shows constant love, and he gives back to a society that literally threw him away. I tease him, telling him he needs to pay back the cost of his medical bills by living to fifty years old. Every day I thank God for not taking him from me.

My second dog, Duke, was originally named Elmo. His is another tragic story with a happy ending. I had asked some friends to connect me to Covenant House, where they were actively doing volunteer work. Covenant House is a privately funded childcare agency that provides shelter and services to homeless and runaway youths. I had created a program there for homeless kids to train homeless dogs. The eventual goal was to hire these kids to work as dog trainers in my facility, as well as in other Covenant House locations.

When I went to a shelter to find the right dog to bring to Covenant House, I brought my close friend Tania with me, to make sure I took only one. At the shelter, I saw Elmo. A German shepherd/Labrador mix, he was handsome with blond hair, perky ears, big brown eyes, and a ready smile. When I saw him walking, I noticed that his back legs were deformed.

X-rays showed that Elmo's legs had been deliberately broken before he had been rescued and brought to the shelter. Dogs like him – handsome with gentle personalities – were usually scooped up immediately at dog adoptions. Elmo had been passed over because no one wanted a two-year-old dog with a disability. He was the perfect dog for Covenant House, and I took him there immediately.

The four kids in my program gathered in a room with Elmo and me, meeting and petting him for the first time. I suggested we change his name, which is something I often do with rescued pets that come from abusive pasts. They can start fresh and respond to a name with no negative associations. Each child wrote a name on a piece of paper. I put the names in a doggy bowl, shook them, and emptied the bowl onto the floor. Elmo walked straight over to the paper with the name 'Duke' written on it, and that was it!

When the program was over, I took Duke home until I could place him permanently. I was happy with Clyde, who loved being an only child, so I really didn't want two dogs. Clydie went everywhere with me – to visit friends and even to the hair salon – and I did not want that to stop. On the one hand, I didn't want to impose on anyone by bringing two dogs instead of one, but on the other, I couldn't just leave Duke at home. I decided to keep Duke and try him out in everyday situations and he behaved very well. Today Clydie has accepted being part of a larger family, and Duke adores his older brother.

Clyde is alpha dog to Duke, but not just because he was with me first or is older. Clyde dominates because of his personality, plain and simple. Dogs all have different personalities. I have seen dog owners with such wonderful first dogs that they get others of the same breed, only to discover that the second dogs are nothing like the first ones. Each dog is an individual, just like every owner is an individual, with his or her own personality, likes, and dislikes. Clyde is a natural-born leader, while Duke is happy to bring up the rear – most of the time.

Duke is particularly good at problem solving, which comes in handy when he wants something that Clyde has. When Clyde has the ball and won't give it up, Duke dashes over to Clyde's bowl and pretends to eat his food. Clyde will immediately drop the ball and rush to assert his ownership over his food bowl, and Duke will take off to confiscate the ball. That is problem solving at its best, and I'm a proud mama – even though it has absolutely nothing to do with me!

Chapter Four

Pain and Pleasure

To this day I know of several well-respected trainers who use unreasonable exhaustion to break a dog's spirit. A common method is to tie dogs to a moving treadmill and force them to run until they are beyond exhausted. At that point all the dog cares about is surviving.

Don't be fooled when a trainer calls a choke chain a 'slip collar'. The name may sound prettier, but it's the same old violent method of choking a dog until he submits to your commands. I have seen these so-called trainers telling owners that spraying Binaca in a dog's face is the way to stop unwanted behavior. Still others physically force dogs onto their backs in what is known as an 'alpha rollover', which is scary and unnecessary. I wish these torturers, disguised as trainers, who are too lazy to actually learn dog behavior, would take to heart the words of Mahatma Gandhi: 'I object to violence because when it appears to do good, the good is only temporary; the evil it does is permanent.'

You have to be smarter than your dog, not tougher. If you think being tougher works, try training a rat to run an agility course. Being tough and dominant will get you nowhere. Compared to working with rats, training a dog is a piece of cake, because he actually *wants* to help, and will forgive you when you are being unclear or inconsistent. If your dog doesn't want to lay down when you ask him to, hurting his neck and shoving him around will only cause his enthusiasm level to plummet even more. I find this abusive behavior utterly unnecessary because teaching your dog to lie down is actually easy – and when you do it properly, he won't even *want* to get up! When you coach your dog using The Loved Dog method, your wishes will become his favorite behavior because he will associate it with great pleasure – not pain.

ALPHA ROLLOVER

There is a training method called the alpha rollover, which a few trainers still swear by. It is suggested that when a dog becomes too aggressive or tries to assume the leadership role in your relationship, you should get down on the ground and use your hands or your body to pin your dog on his back so he can't move under your weight. Then, from that terrifying position, the trainer is meant to stare the dog down into submission.

Does it work? Ask my friend Michelle, a Canadian woman whose Jack Russell terrier stiffened up whenever she got close to his food bowl. Instead of systematically coaching her dog to accept her presence (as I describe in chapter 8), she followed a trainer's advice to do the alpha rollover and pin her dog to the ground when he growled at her. She called me when she got back from the emergency room after her terrified dog had bitten her.

People tell me that the alpha rollover is natural, and that an alpha wolf or dog performs this technique to create submission in other pack members. That is absolutely not true! You may see a snapshot of a submissive wolf lying on his back in front of a higher-ranking wolf. But if you'd seen the sequence of events that took place beforehand, you'd realize that nobody put the wolf in that submissive posture. He got there on his own, whether he was playing, or looking for acceptance from his leader.

DISPELLING MYTHS

Training dogs without abuse and pain requires an owner to better understand their underlying nature and motivations. In order to do this, we must dispel two popular myths. The first is that dogs don't have feelings. However elementary this may sound, you would be shocked to know how many people don't realize that dogs have a complete range of emotions, which are very similar to those of toddlers. Dogs feel loss, joy, and disappointment; they get obsessed; and they love surprises and games. Although they are quick to forgive, they will remember things. Every time Clyde hears a motorcycle, his ears go down and he runs to hide in a corner. It's obvious that something in his past taught him that the sight and sound of motorcycles are negative or dangerous.

The second myth is that dogs will do anything to please their owners. The truth is that dogs will do anything to please *themselves*. If these things also please their owners, it's just icing on the cake. After working with thousands of dogs throughout many years, I have seen that the driving force behind any dog's behavior is an association with pain and pleasure. If you understand and align your needs with his needs, then he *will* do anything to please you. Just like humans, a dog will always seek pleasure and try to avoid pain.

Sometimes, the divide between what humans and dogs find pleasurable is vast. You wouldn't roll against a dead animal's rotting corpse for any amount of money, right? But to a dog, a smelly carcass is like Chanel No. 5. This seems impossible (and disgusting), but the truth is, she's having a fabulous time, and stopping feels like a bummer. I'm not suggesting you allow your dog to roll around in manure and animal cadavers just because she likes it. I am suggesting that if you want her to stop what she considers a pleasurable behavior, you'll have to find something even more enticing with which to distract her. It's up to you as a responsible dog owner to understand her needs and desires as a dog, and to provide her with healthy outlets.

Oprah Winfrey's golden retrievers are three of my favorite dogs. I met them when they were three months old, right after Oprah took them into her extended family, and I have been privileged to coach and play train them through their various stages of life. One of them, Layla, is a great adventurer who could have been the leader of Outward Bound, if she was human. I used to call her Layla Cousteau because whenever she could, she'd dive into the pond with her eyes open, looking for fish! She was in her own private world, and when someone tried to get her out of the water, it was as if she suddenly went deaf. In the pond, she was in doggy heaven, and no amount of begging, pleading, or threats would reach her. When I saw the extent of her diving and exploring obsession, I knew that if left to her own devices, she would stay in there for days, which was frustrating for Oprah, and also not particularly good for Layla's health.

When I play train a dog, I always try the easiest method first, or the path of least resistance. If that doesn't work I'll try other ideas. Since Layla felt that getting out of the pond was a bummer, I had to teach her to *want* to get out. I started by putting a chicken treat (the gold medal of treats) in my hand, took Layla by the leash, and walked her to the pond, feeding her treats and saying, 'take it, take it, take it', as she happily munched away. Then I took a liver treat (silver medal) and threw it into the pond. 'Dive, dive,' I told Layla. She did not need to be asked twice! She flew into the water to retrieve the treat.

After she swam around for a few minutes, I dangled a gold treat from my fingers and lured her with it by saying, 'Take it, take it.' Layla hesitated. I could see her mind churning, *What should I do? I really like those treats.* She compromised and took a step toward the edge of the pond. When she stepped out of the water, I gave her a jackpot of gold treats that I reinforced with the words 'Take it, take it', using my happiest voice to praise her, as if she had just won the World Cup. Then, to Layla's surprise, I tossed a silver treat into the center of the pond, called out 'dive, dive' and she rushed back into the pond. That day she went in and out of the pond whenever I asked her to – with the help of some wonderful treats, my happy tone of voice saying 'take it' and my arms extended wide open as a big reward. On a scale of one to ten, getting a jackpot of gold treats and then being allowed back in the pond was an eleven! Layla's pond explorations are no longer a problem, since she happily comes out when asked.

What I did with Layla was an exercise in healthy behavior

modification that served us both. My challenge was to make her understand that when she got out of the pond, the fun would not stop. With my method, coming onto land was not a bummer, but a gift. Layla finally understood that getting out of the pond was not a punishment or the end of her playtime. I could see her brain registering it, and then she made the change herself because she wanted to, not because she was afraid. I simply aligned her pleasure association with what *I* wanted, and she started getting out of the water when she was asked to do so.

One morning I was called to the lobby of The Loved Dog to meet a female Rottweiler named Cherokee. Her owners, a lovely couple, told me she was afraid of everything, and I only had to see that she was missing a toe to understand why. She was a victim of an abusive trainer who used electrical currents on her toes and would only stop when she obeyed. By the time she was finally rescued by this wonderful family, her toe required amputation since the nerves had been ruined. She had also spent most of her days confined to a cage in a 'training' facility, another form of abuse. My heart went out to Cherokee, and I began to coach her myself. Although Cherokee was not aggressive, her situation was more urgent because she was a Rottweiler, a breed that often is misunderstood in a negative way.

Have you ever felt uncomfortable at a cocktail party until you struck up a conversation with someone fun? I knew that having a friend would help Cherokee get through the process, but she was too frightened to approach another dog and make friends the way most dogs naturally do. I needed to get creative. After I showered Cherokee with love and she trusted me, I moved on to the next step of Operation

Cherokee. I smeared honey around her mouth and introduced her to Ginger Bosley, an easy-going dog I thought she would really like.

When the two dogs first met, Ginger crouched on the floor and lay in a submissive posture, which let Cherokee know that she was not a threat. Cherokee gave a little growl and began to check out Ginger. When Ginger smelled the honey, she started licking Cherokee's mouth, and they were instant friends! I felt like a proud 'doggy matchmaker'. With the support of a good friend and non-violent coaching, Cherokee began to heal. Her fear dissipated, and the two dogs shared a fantastic friendship for many years.

In the cases of both Layla and Cherokee, I needed to figure out what made each dog tick before I could bond with them and coach them. Each dog's fears and desires are unique to their personalities, but we also need to be aware of the seven basic needs that all dogs share.

Chapter Five

The Seven Basic Needs

It is important to understand that a *need* is just that – a need, not a preference or a desire. Needs are something that a dog cannot do without, or the quality of his or her life – mentally, emotionally, and physically – will deteriorate. Through many years of working with dogs and their families to solve or prevent problems, I've discovered that whatever their sizes, whether they are apartment dwellers or live in a house with a huge backyard, however different their daily routines may be, all dogs across the world have the same seven fundamental needs. A beagle may like to sniff more than a shih tzu, and a Labrador may like to play more than a Chinese shar-pei, but their basic needs are the same.

Dogs revert to what we call 'problem behavior' when even one of their needs is not being met. These seven needs are interconnected and are required to create a whole, healthy animal. If any one essential part is missing, the system will break down, no matter how well you maintain the rest of them.

The seven basic needs of a dog, not in order of importance, are:

1. **SENSE OF SECURITY**
2. **COMPANIONSHIP**
3. **UNDERSTANDING THE HIERARCHY**
4. **SURPRISES/EXCITEMENT**
5. **FOOD AND EXERCISE**
6. **MENTAL STIMULATION**
7. **LOVE AND CONNECTION**

1. SENSE OF SECURITY

Like people, dogs crave security, which lies at the core of their survival. As much as we rely on the assumption that the floor will not disappear from under our feet, your dog needs to count on you. He needs to know that when he comes to you, he will be praised and given a treat, a toy, or some love.

Your dog experiences profound security through having a steady routine – his daily activities, the way his family treats him, and the way he interacts with the world. Only if he can draw from a deep well of security will he feel comfortable trying new things, and all training is ultimately the introduction of new experiences and behaviors.

When an owner's behavior is unpredictable, a dog will become confused. If you call him and when he comes, you smack him for something he did ten minutes ago, he will not understand your behavior, and he will not be so keen to come the next time. From then on, whenever he hears his name called, he'll pause, look over to assess his owner's frame of mind – and oftentimes, he'll choose to go in the *opposite* direction.

Before we can relax and enjoy our lives, most of us need structure and certainty that our most basic needs will be met. If you have ever worried that you might lose your job, you know how impossible it is to focus on anything else. Now, imagine if you were in such a state of uncertainty – not for a week or a month, but your whole life. Such is the lot of a dog who is denied a steady routine. Given the importance of predictability in a dog's life, it is likely that introducing a new request or exercise will initially cause him confusion. In fact, whenever you see your dog suddenly having a behavioral problem, it almost always relates to a change being made in his routine without clear communication.

I recently helped Melissa teach agility to her sweet Rottweiler mix, Spencer. At The Loved Dog, we have an agility course where dogs can have fun with their owners while they play on a doggy jungle gym. Part of the course is a wooden A-frame structure that is about four feet wide, with stairs that go up on one side and down the other. The dog climbs up the stairs on one side, and when he reaches the top, he scurries back down the other side. When he gets about one foot from the ground, we give him the signal to stop. This is where extra control comes in, and doing it successfully earns treats, praise, and points for the dog in an agility competition.

Spencer had always been a champ at agility, and Melissa had successfully taught him the emergency stop. One day, though, I noticed that Spencer refused to use the A-frame equipment. What had changed? Melissa told me she had entered Spencer in an agility competition over the weekend. That in itself was fine, but it was not fine that the trainer had insisted that Melissa use his stress-based method to signal Spencer to stop. The trainer wanted Melissa to hold up her hand and to yell, 'Stop! Stop!' in a loud, commanding voice as she body blocked him. In an attempt to please the trainer, Melissa had given in and started using this harsher voice and attitude.

When she tried to suddenly stop ninety pounds of happy, running dog by using stress and intimidating body language, it backfired. Spencer was unsuccessful and confused, and he refused to get back on the equipment that he had loved so much. When I watched him try it at The Loved Dog, I could see his eyes pleading with his mom, *Do you want me to come down or not? Please help me understand you.*

I took Melissa aside and said, 'Let's forget the A-frame and work with the stop exercise on the floor first. Once he remembers the pattern, you'll be able to stop him anywhere.' In record time, Spencer was back on the A-frame. He happily climbed up and down, and he stopped whenever Melissa asked him to do so – with a hand signal and the gentle tone of voice to which he was accustomed. He got a lot of love and treats when he did it right, which gave him the confidence he needed to succeed every time.

2. COMPANIONSHIP

Dogs are naturally pack animals that need to be social. Wired like her wolf ancestors, a dog likes to live in cooperation with others, which ensures a good chance of survival in nature. In the wild, wolves lie around on top of one another, enjoying the physical closeness. Even if you have a huge backyard, have you ever noticed how quickly your dog wants to come back inside to be with you? However large your home is, your dog will always be a few inches away. Dogs possess an innate desire to feel the security of being part of a pack and knowing that someone is there to watch their backs.

Some owners think that because they work at home, their dogs get all the companionship they need. This is not usually true. Would your husband feel fulfilled at the end of a workday if you gave him a peck on the cheek, put his dinner on the table, and went back to your computer to keep working late into the night? Psychologists stress that you need to spend quality time with your spouse and your children, and you also need to give your dog quality time. Companionship means interaction, and it requires your active attention, energy, and commitment.

Throughout the day your dog needs active feedback. She needs your attention, with your tone of voice reinforcing the words and actions that she knows. Smile often and let her know she's doing a good 'down', a good 'fetch', etc. It only takes a second, and it reinforces the connection between the two of you, assuring her that you are there.

Last week I was outside, tossing a ball for Clyde, when I saw a woman walking down the street with her Lhasa Apso. He was doing what dogs do, checking his 'p-mail' and taking inventory of everyone on the street. Meanwhile his owner was reading a book while the leash dangled from her other hand. These two were merely coexisting. She could not have looked more disinterested in her dog. When a toddler is about to go down the slide, he calls out, 'Mom, watch me.' He wants to know she is there, that she is connected to him and proud of him – and her undivided attention reassures him. Dogs are no different.

3. UNDERSTANDING THE HIERARCHY

Like a wolf, a dog must always know who his leader is. Without a clearly designated leader to show the way, a dog will become nervous and will attempt to take things into his own paws. Through his leader a dog finds his place in the pack and learns how to function in life. You should always be seen as the alpha, or leader – there must be no question about that! In order to understand this better, let's take a look at a wolf's rules and value system.

Wolves must have clear rules to keep the peace with so many sharp-toothed pack members looking for ways to assert themselves. One important rule is that the alpha wolf *always* eats first. When I ask a client if the dog is eating on a schedule and they say yes, with pride, I ask them to describe their breakfast routine. They usually give breakfast to the dog to get it out of the way, feed the kids, and if they're lucky, they grab a cup of coffee before they leave the house. This kind of behavior sends a message to your dog that he is the leader. To make it clear, particularly at the beginning of your relationship with your dog, make sure that you eat first, even if it's just a cup of coffee. Put his food on the counter and give it to him only after you put the dishes away.

The second rule is that the alpha wolf gets the highest and best sleeping area. Your dog may try to take up all the space on your bed and even try to kick you off the couch. If you want her on your bed, she has to know she is being invited – a privilege that only you can give her. After all, it's *your* bed. You can invite her to snuggle with you, but don't share your sleeping space until she's clear that you are her leader.

Third, the alpha wolf must be the best hunter, which he demonstrates by always winning chasing, wrestling, and tug-of-war games that practice hunting skills, which we'll learn in chapter 11. Over time, as your dog cherishes you as his leader, you can loosen up the rules.

When there are two or more dogs in a family, an extended hierarchy should form in this order: you, highest-ranking dog, other dogs. Both wolves and dogs try to improve their pecking order. But what would they do with a battered pack member when everyone is needed for the hunt? Instead, they play, which allows them to assess one another's skills, just like we humans do when we play sports. Let the dogs establish their position among themselves without interference from you – unless it ever escalates into a full-fledged fight.

It's important to understand that all dogs in the same family are *not* equal. They are not meant to be. In this way, dogs differ greatly from humans, and we need to respect their wolflike value system, even though it is different from our own. Although you would never do this with your children, your job is to accept the hierarchy as *they* establish it. It may be difficult for you if a new dog is taking a leadership role over your older dog. However, you must give more privileges, not less, to the dog that wins the highest position. Human beings love an underdog, and it may feel like you should pay extra attention to the lowest dog on the totem pole – but that will only cause unrest and could lead to fights among the pack. You must behave counterintuitively in this way, because in a happy pack, everyone knows where he belongs and what role he plays.

I often see people reprimanding an older dog when he growls at the annoying new puppy that just joined his household. This is a mistake. The older dog should get a bonus for helping the newcomer understand the pecking order. When you support the lower-ranking one, you are actually enabling more disputes to come, and it confuses the issue of who belongs where.

After I got a desperate call to take a Dalmatian named Rocks and find him a new home, he temporarily joined my family. My pack went from two to three males. In our pack, Clyde is the undisputed second after me, and Duke and Rocks had to 'duke it out' for third in command. As much as I wanted to help Rocks during the process, I had to let the natural order take its course. When Duke sat at the top of the stairs and put on an ugly face each time Rocks wanted to go up, it quickly became evident that Duke was Number Three and Rocks was Number Four. Duke blocked his passage and took toys away from him, and when Rocks tried to climb up on my bed, uninvited, I did not baby him. I wanted to, but I knew that giving privileges to a lower pack member over the others would potentially create a fight. We simply have to respect the dogs' natural order, even if it does not resonate with us.

4. SURPRISES/EXCITEMENT

While your dog needs predictability and routine, he also needs the opposite: some surprises and excitement. A large percentage of behavioral problems stem from the dog simply being bored and his wolf nature will come up with some primal and creative remedies, such as chewing, barking, and digging. If you leave your dog to entertain himself, you'll be amazed at how resourceful he can become. If he doesn't run away to look for adventure elsewhere, he may become an interior decorator, a landscaper, or neighborhood news broadcaster.

Other popular self-appointed dog jobs are overzealous bouncer, ID sniffer (of guests' crotches), archaeologist, and demolitionist.

A good rule of thumb is that your dog needs to have at least one surprise a day. This can be a walk to check his 'p-mail', a car ride, a playdate, or a visit to the doggy park. Get a bunch of interesting toys, treats, and other rewards for your dog, and dole them out when least expected. Happy surprises are a great motivator!

As the leader, how much excitement can you generate in your dog? You don't want to be a boring or, even worse, a feared owner. When I see a dog trainer not allowing a dog to show excitement (they might call him 'hyper') I determine that he doesn't really know how to train dogs. These charlatans use Binaca spray, choke chains, and other horrendous methods to prevent the dog from getting excited, since they don't have the skill to teach or manage a rambunctious dog to use his energy as part of the learning process. Following The Loved Dog method, I want you, the leader, to generate excitement in your dog. He should be thrilled to learn new things! The more your dog wants to be with you and learn from you, the easier it will be to coach him.

5. FOOD AND EXERCISE

Dogs need exercise in order to blow off steam and become relaxed and compliant. It's also a fabulous opportunity for you to bond with your dog and get some exercise yourself. We bond deeply with people who enjoy the same hobbies and activities, and dogs are no different. When I don't have time for a full exercise session, I like to encourage them to play fetch. The dogs let out some steam, and I don't even have to change my outfit since they do all the running. Just do whatever will tap into the dog's enormous energy reserve. Usually, the younger the dog, the more exercise she needs. Her breed will help determine how much she requires, but all dogs need exercise to make them happy pets that can be coached successfully.

I once worked with a three-year-old male Labrador named Perry, whose owner, Charles, was a really great guy, a TV sports commentator. Charles wanted me to help him train Perry to stop jumping on guests. When I arrived, I met Perry – the bouncer – at the door. He checked my ID (sniffed my crotch) and jumped all over me, hoping for some attention and excitement. I quickly demonstrated to Charles how to distract him from jumping, using a technique I'll describe in detail in the coaching section. Within a few minutes of coaching, Perry stopped jumping. But as I watched the young dog straining to control himself, as if he were a pressure cooker about to explode, I realized that Perry's real problem was insufficient physical exercise.

In Charles's mind, he was constantly giving Perry love and attention, because he worked at home and took the dog for a twenty-minute walk every day. That might have been enough for an older dog or a different breed, but twenty minutes of walking did not even scratch the surface of Perry's exercise needs. Labs are bred to hunt birds and do field work for long periods of time. Since his daily walk wasn't enough to satisfy him, Perry had given himself the job of frisking whoever walked through the front door.

Remember that your dog does not look at exercise like you do. You may hate going to the gym, but your dog looks at exercise as a reward. She wants to let out her wolf nature, running, jumping, playing, and having a wonderful time. One thing is for sure – without exercise, don't expect your dog to come when you call her, or at least don't expect her to return to you (and her boring, predictable life) quickly.

6. MENTAL STIMULATION

Did you know that a dog is capable of learning up to one hundred and fifty different words? I am constantly baffled when I discover that an owner has not taught his dog a word for going to the bathroom. We teach children a word for it, in order to show them the proper behavior, so why not have a word for it with the dog? Not only will she understand you better, but you'll be able to tell your dog to 'drink', then 'go potty', and then 'go in' to your car and take a road trip. Her brilliance will amaze you!

I suggest you make a list of all the ordinary things she already does – things you don't usually pay attention to – and give them all names. When she drinks, say 'drink, drink'. When you take her to the park, say 'park!' When she sticks her head out of the car window, repeat, 'window, window, window'. Make a list of new things you can teach your dog to keep her mind stimulated, with a name for each new behavior.

Mental stimulation is the reason your dog loves to sniff the ground when she walks. The astounding array of smells that our dogs can detect is an information super-highway, and her daily walk is the equivalent of logging on to cyberspace. At home, she can use her complex brain to learn the difference between the *LA Times* and the *Wall Street Journal* (they are wrapped differently) as well as how to bring you one or the other upon request. Ask your dog to bring your tennis shoes, and then ask her to get the flip-flops. It may take time and effort, but I promise you that with a little patience on your part, your dog is capable of unbelievable tasks.

My friend taught his four-month-old golden retriever, Mickey, to dive into the pool to retrieve his golf balls. Another client taught her dog to pick up her toys and drop them in the toy basket when the game was over. Believe it or not, I actually met a dog who could open the refrigerator and bring his owner a soda! The sky is the limit!

Changing your dog's usual form of exercise or play-training routine will help to stimulate her brain. If your dog is accustomed to lying down after she sits, ask her to fetch the newspaper instead. At first, she'll be surprised at the change, but then she will expect change in the future – and end up more responsive to your requests. Instead of simply giving your dog the new toy you just bought, ask her to stay while you hide it somewhere. Let her nose be her guide, something that she loves to use but is rarely asked to. Changing locations for play-training offers additional stimulation. Any dog can learn a new behavior in one place, but the trick is to teach the dog to adapt and perform wherever she is. Search-and-rescue dogs are coached to do their work in any environment – so take your training outside, to a friend's house, or to a different yard or park.

Think of your dog as a sponge, willing to take everything in and eager to learn new things. She loves the adoration she gets from people when she does tricks and learns new words. Each day, teach your dog to play hide-and-seek or to learn your family members by name. A dog is a living creature with a hungry mind and a great capacity for love and gratitude – as long as you keep her interested and excited about life.

Let me caution you about getting a smart dog, like a Border collie, whose intellect and concentration levels are superior to most other dogs. (In some cases, superior to their owners!) A smart dog will learn how to manipulate you as successfully as a smart kid can, and their intelligence can quickly shift to learning destructive behaviors if they don't get constant attention or mental stimulation. They can open locked doors, find the hidden chicken bones in the garbage, and move objects from shelves to their own special hiding places. They'll get you to do things their way, and you won't even realize you're doing it! But if you can commit that extra time, attention, and money, go for it. The rewards for both of you will be immense.

7. LOVE AND CONNECTION

Dogs need to feel connected to their owners and to their lives, and it all starts with you, the leader of the pack. It is not enough just to exercise your dog, give her some food, and then put her away like a toy until you feel like taking her out again. Similar to a toddler, a dog needs what psychologists call 'secure attachment', an invisible cord of love that connects one being to another. They need to know that even with some physical distance, you are still there for them, always connected. This is a kind of connection that goes beyond words, and it is important not to underestimate that connection in a dog's life.

Clyde and I once went on a hike with a friend Evan and his two-year-old collie, Sunshine. We got into a discussion about the difference between connection and love, and I pointed out that I believed Evan's dog absolutely loved him. But from what I observed, I did not think Sunshine felt *connected* to him. I asked Evan if he was willing to do an experiment, and he agreed. We walked for a while, and when we were out of our dogs' sight for a moment, I hid behind a tree. Evan and Sunshine kept walking, but after a couple of steps, when Clyde noticed that he couldn't see me, he stopped walking and began to search for me until he located me behind the tree. I praised him enthusiastically, 'You found me, yeah!' Then we kept walking.

A few moments later, when we saw that the dogs had stopped a moment to sniff something, Evan hid behind a tree, while Clyde and I walked on – and so did Sunshine. She walked on, her tail wagging with abandon, oblivious to the fact that Evan was nowhere to be seen. After a while, I had to physically lead Sunshine to Evan. Although Sunshine loved Evan, they were not really connected. Sunshine did not feel like she was missing someone when Evan was not in her sights. Evan was stunned, since he felt that he gave his dog so much love. That was true, but love is not all that a dog needs.

Before you get a dog, you should ask yourself why you want one in the first place. Is it for you, your children, your other dog, or all of the above? Remember that getting a new puppy can be difficult and confusing, and it takes a bit of adjusting to understand your role as caretaker and leader. If you are willing to fill your dog's most basic needs, and can take the time to learn who he is and to coach him with love and clarity, you and your dog will thrive happily together.

Part Two

Play-Training

Chapter Six

A Common Language

Finding a common language with your dog might sound like mission impossible, but it isn't. Don't buy the idea that only certain trainers have the rare gift of understanding dogs – you have the same gift. After all, dogs are not aliens from outer space. They actually are pretty easy to understand if you try to look at things from the dog's point of view. Once you learn how they think and communicate, you can help your dog fit perfectly into our human society.

With patience, compassion, commitment, and the right tools and techniques, you'll be amazed to see that your dog is like a window to the natural world living right beside you. We can see it in their habit of making circles before they lie down on their beds, how they try to bury a toy in the couch, move their noses as if to cover their prize with dirt, and toss their toys in the air like they were captured prey. Dogs are truly nature's ambassadors. And while humans communicate mostly through spoken language, dogs communicate through body language, facial expressions, and sounds. That means we should be very careful what messages we are sending, even subconsciously, to our dogs, who can sense even the subtlest of signals.

There must be a consistent language within the household as well as agreement over which doggy behaviors are acceptable. Should the dog be invited on the couch? On the bed? Should he be allowed to jump on you as a greeting? What kind of rewards and chew toys are acceptable? Do you use the same names and signals for particular behaviors, and do you expect the same results? Has your dog figured out whom he can best manipulate? Think about how confusing it is when Mom uses a certain signal for 'sit', and Dad uses the same signal for 'down'. How will your poor dog determine which signal is for which command?

I once worked with a dog whose owner lived with two roommates, and anything the dog learned was undone by the others in the house. In another case, a dog incessantly barked for attention, and despite his owner's efforts to train the dog, his wife just declared, 'I can't take it anymore', and she let the dog have his way. Neither of these scenarios will end up in success. In order to teach your dog good manners, remember: if a behavior is rewarded, it is likely to increase. If a behavior is not rewarded, it is likely to decrease.

It's just that simple. Be aware of the messages you're sending, and the entire household must agree. There really is no right and wrong here because with a dog, there is no 'one size fits all'. It's up to *you* whether or not the dog will sleep on your bed or if he's allowed to walk all over the coffee table. Just be consistent in what boundaries you set.

BE AN OPEN BOOK

Dogs can read us as expertly as wolves can read body language during a hunt. When you go to a certain closet to get your jacket, your dog jumps with joy because it means you're both going for a walk. When you grab your keys, the dog makes the connection that he is going for a ride. I chuckle to myself when a dog owner boasts that his dog has *graduated* from voice commands to hand signals. The truth is that a hand signal is easier for a dog to understand. In fact, in many exercises, it is only after I teach a behavior and set a pattern with a hand signal that I give it a name, like 'sit' or 'down'.

Because your dog needs to read your face when you coach him, you should avoid wearing sunglasses, which prevent eye contact. Your dog needs to be perfectly clear as to whether or not he is pleasing you, and giving him *exaggerated* feedback will help. Build a pattern in your dog's mind, focusing on the positive. This may require retraining yourself, since many people are in the habit of ignoring good behavior and only paying lots of attention when something is not working. Always remember that you and your dog are a team, and your job is to be the clearest team leader possible.

But how do we communicate with our dogs who are a completely different species from us? It's actually easy. We simply need to find a common language that includes words, body language, facial expressions, exercises, and games that are familiar to both humans and dogs.

In my training DVD, there is a boxer named Henzo who was, for the most part, only fluent in his native wolf language. Henzo, desperate to communicate with his owner, Patricia, was trying to engage her in a game of chase by jumping up and stealing food from the kitchen counters. He was not trying to drive her crazy nor was he a thief nor was he starving to death. He just hadn't been taught a language that he and Patricia both understood, and he was trying to get her attention the only way he knew how – using wolf games.

His habit of pulling on Patricia's robe when she first walked into the kitchen in the morning was another attempt to get her to 'talk' to him. It was an invitation to play tug-of-war. Patricia told him no over and over again, and decided he was being obnoxious. When I met Henzo, I said, 'Your dog is not obnoxious. He's trying to tell you something, and that's the only way he knows how.' What could be more upsetting than to be called bad for simply trying to make a connection?

TEACH YOUR DOG ENGLISH

Finding a common language means that both you and your dog must learn each other's communication styles. And just as I am asking you to learn to understand wolf, you can ask your dog to learn English! To begin, I recommend you take inventory of everything he's already doing right. Even if he's badly behaved, he is probably still doing many of the things you ask him to, like sitting, lying down, or playing with a toy. Acknowledge his good behavior by repeating the word for the action over and over, such as 'sit, sit, sit' when he's already sitting. If your dog is already lying down, say 'down' with an approving voice. When he's playing with a toy, say 'play, play, play'. When he does that long and luxurious stretch in the morning, repeat the word 'bow' in an enthusiastic tone. Even if he's yawning, you can say 'yawn, yawn'. Anything the dog does right, verbally acknowledge the behavior. When you say a word like 'sit' to your already seated dog and give him a treat, he'll perk up. *What's this?* he'll think. *I'm getting a goody for just being me!* With the ability to please you, he'll see himself as being brilliant, and he'll want to learn more.

There are many benefits to teaching English to your dog. If I'm on my way to a friend's house with Clyde, I ask him to relieve himself outside first with a 'Hurry up', and he happily waters the nearby bushes. When we go into my friend's house, he may sniff around, but I don't worry that he might pee because he just did it outside. I call his bathroom business 'hurry up' because that's how I feel about it, but any term is fine if it's simple. Some people say 'Do your business' or 'Go potty'. Oprah's term is 'poody' and one of my Yiddish-speaking clients uses the word 'pishen'. The other day, I heard someone say '*baño*'. The different names crack me up. Call the action whatever you like, as long as you establish a word that makes sense to both of you.

Once you've taught your dog a vocabulary, you can tell him what you'd like him to do by using a simple word. When I was filming my DVD, Clyde and Duke sat patiently in the back of the car until I said the word 'window'. Off they went to the window – and toward the camera – because they knew exactly what it was and what I wanted them to do. When I praised them, they looked like they were about to burst with pride! You should see beach bum Clyde when I say the word 'Malibu'. He perks up like someone just gave him a huge jackpot of gold treats!

USE SHORT WORDS AND PHRASES

A common mistake is to use a lot of different words when asking a dog for something. I hear people say, 'Come on, good boy. Come here and sit down. No, no, sit. Come over here.' Your dog just heard 'come', 'sit', 'down', 'no', and a bunch of words he doesn't recognize. He's thinking: *What the heck does my owner want?* The fewer words you use at any one time, the better the chances that your dog will understand you. I suggest one word per behavior, the shorter the better, or all the dog will hear is 'Blah, blah, Fido, blah, blah.'

Constantly saying 'good boy' or 'good dog' does nothing to hone your dog's ability to understand. If he doesn't know which of his behaviors is good, how can he repeat the behavior and succeed again? When you want to praise your dog, try attaching the word 'good' to the behavior you'd like to see, such as 'good sit' or 'good down'. Better still, leave out the word 'good' altogether and use a singsong voice to stretch out the word 'sit' making it sound happy and fun. Even if at first you feel weird 'singing' words and using baby talk to your dog in public, do it anyway. (You can blame it on me!)

Even Henzo's owner, Patricia, a highly educated therapist whose work is based upon the nuances of communication, forgot everything she knew when it came to her dog. She had a hard time remembering to use the word 'good' without adding 'boy'. Habits are not easy to change, and she'd been putting the two words together for a long time. It took her a while to retrain her mind. I explained that she was acknowledging the action, not the dog, and once Henzo began to respond to her simpler communications, Patricia's joy in reaching him helped reinforce her new way of talking.

Watching herself and Henzo in my DVD did a great deal to help Patricia change her communication style. Patricia was stunned to see that she continued saying the words 'good boy' immediately after I asked her not to. If you're having trouble changing your patterns, ask someone to watch you work with your dog, or ask them to videotape you training him. After a few reminders you'll get the hang of it and so will your dog!

SILENCE

The opposite of repetition is silence, and a dog needs some of both to learn our language. Have you ever tried to learn something new when someone was talking to you non-stop? It doesn't work!

I recently placed Rocks, my Dalmation friend, with a very smart lawyer, and showed the owner to teach the dog to sit. First I used a hand signal – again, the easiest way for any dog to learn, since wolves use body language. Then I said 'sit', and I waited a few moments to see if he understood.

Rocks sat easily and quickly, but when the new owner took over and said, 'sit, sit, sit, sit, sit', Rocks looked confused. He already knew the word 'sit', but since it was being fired at him repeatedly from a stranger with a different tone of voice, he needed a little time to process the information. I told his new owner that when he was saying a familiar word to Rocks, he should say it once and then go silent for a short time. This would help Rocks recognize the word in his new surroundings.

When a dog gets the opportunity to process something with his own mind, he has a much better chance of remembering it. If you were working out a math problem in school and someone gave you the answer, you *might* remember it later. But if you'd come to it on your own, you would most likely not forget it. We don't remember anything as well as we remember our breakthroughs, and it's the same with your dog.

A friend, Scott, has a golden retriever named Mickey. Scott is an enthusiastic guy who taught Mickey a lot in a very short period of time. Mickey was eager to learn, since he felt so much enthusiasm coming from his owner. He especially loved to roll over, which he learned fast. One time, instead of telling him to roll over, Scott suddenly told Mickey, 'Off!'

He wanted his dog to stop rolling over and to back off instead. Without thinking, Mickey rolled over. It didn't matter how many times Scott repeated the word 'off', Mickey had fallen in love with rolling over. He had been trained so quickly, he went on automatic pilot, and rolling over was the only thing he wanted to do. It was as if Scott needed to push Mickey's reset button. He slowed down a little, and Mickey was so smart, he caught up. Pretty soon, Mickey was backing up the entire length of Scott's large backyard. I swear, he would have walked backward all the way to China if Scott had asked him to!

When dogs pick up things very quickly, they often lose them just as quickly. Given a little more time to figure things out without pressure or aggression, they will remember because they worked it out themselves. Then you'll have a dog who listens, and his success will send his enthusiasm level for training off the charts!

Allowing time for your dog to process information is a sign of respect. Show your dog a desired behavior, give him a word for it, and give him some time to think it through. If he doesn't get it, don't correct him. Just wait patiently until you see the light go on in his head. For example, when you teach a child to recognize the color blue, you point to the color and say the word 'blue' once or twice. Then you let him process it. There's no mad dash to make him immediately understand. The child eventually recognizes blue with no pressure from you. Similarly a dog is an individual with his or her own timing, which must be respected.

TONE OF VOICE

In dog training I use four levels of tone of voice: normal, excited, unhappy, and urgent. I consciously play my voice like an instrument, using these different levels with my dogs at appropriate times, offering them constant feedback. Within the levels, there is also a range. Excited is not always the same, as it can range from pleased to 'My fabulous dog and I just won the lottery!' Neutral is just a resting place while negative feedback should range from Level One, in which the dog made a simple mistake and you calmly ignored him, all the way to what I call Level Eight, which is my most displeased tone.

I'm always looking for the opportunity to go to the happy tone of voice and 'make a party', which is what I call making a fuss so dogs can feel good about themselves. I generally talk to them in a normal tone of voice, but I like to give them the happy or the unhappy tone as immediate feedback to guide them along. I rarely use a Level Eight tone of voice, unless the dog is putting himself in danger. Just as if your child reached out to touch a hot burner on your stove, there are situations in which you can't and shouldn't speak calmly. You have to make sure she associates getting too close to the fire with a negative experience that she will have no desire to repeat. This is what wolves do with their young. When the alpha wolf needs to send a signal, he starts with a little growl. Then he shows some gum, and finally he bares all of his teeth (or 'visits the dentist'). The wolf is saying to his pack members, 'I don't want to hurt you, but I will if I have to.' And you can be sure the threat alone works wonders.

When I use Level Eight, it looks so scary, I have to warn people in advance that I'll be acting. Have you ever noticed that when animals are getting ready to attack, they make themselves seem bigger, hoping their opponent will give up without a fight? The hair on the back of their necks goes up, their ears become erect, and their tails are stiff and carried as high as possible – all so they will look more threatening. Such a display helps avoid physical conflict, and I adopted my own Level Eight display so I would never have to use physical force against a dog.

Here's how it looks. I take a deep breath in and hold it, taking only shallow breaths when necessary. I make a super-angry expression, lift my shoulders, and stick my face out. I speak forcefully in a low-pitched tone of voice, and I stiffen my body – a wolf signal that I may hurt anyone who continues to cross me. I explain that while Level Eight may look weird and is certainly intense, it's the opposite of yanking the dog and hurting his neck with the protruding metal pins of the prong collar.

I always choose a scary tone of voice over physically hurting a dog. If only people could understand that it's okay to momentarily sacrifice their images and look weird to spare harming their dogs! Then, once you have successfully used Level Eight, all you need to do later is take a loud deep breath and give your dog 'the Look'. He'll think to himself, *Well, never mind. I think I'll listen to Mom before she goes all 'wolf' on me.*

Some dog owners have a really hard time sounding enthusiastic, while others can't seem to get the panic out of their voices. Neither of these extremes will work since dogs need to hear an appropriate range of emotions. When CJ, a chocolate

Lab mix, backed off successfully for the first time, I 'made a party', praising him and acting like he'd just won a gold medal. But his owner, a sweet, shy young woman, smiled and quietly told the dog that he'd done well.

'You need to get excited!' I told her with enthusiasm. 'How else will CJ know he accomplished something fantastic? Sing out!'

CJ needed to know that his accomplishment was worthy of a whole lot of hooting and hollering. Then he would want to keep training. I advised his owner to drop the opera glasses, take off her gloves, and start throwing around some vocal confetti.

A friend of mine recently admitted to me that she never knew how sternly she spoke. As she became aware of how she sounded to her dog, she realized that she was using the same tone of voice with her young daughter. When she changed her tone of voice, her dog began to do whatever she asked – and the difference in her *daughter's* behavior was just as remarkable.

On the opposite side of the coin, I worked with a woman named Morgan, a gentle soul who practiced meditation. She felt uncomfortable using her voice to impart anything negative, which created an out-of-balance training for her dog. I had to impress upon Morgan that we live in a world of duality, black and white, clear and gray skies, and we each have our likes and dislikes. As much as we want to give our dogs positive reinforcement, there are times when we must use negative feedback in an appropriate form. When Morgan's dog refused to listen to her calm requests, I encouraged her to breathe in hard, tighten her shoulders, and use a tone of voice that her dog could not mistake for appreciation or approval. How else would her dog learn what was acceptable behavior and what was not? I still have Morgan's e-mail in which she said something like, 'It's still difficult for me to use a full range of emotions, but I'm doing it, and it's working.'

I was compelled to use Level Eight during a training session with Luke, one of Oprah's golden retrievers. I was told that Luke saw a dog walking outside the property gates, and he ran across the street to play. It was natural that he was attracted to other dogs, but running across the street could have gotten him killed. My performance would require acting skills and energetic responses that knew no bounds.

Since Luke knew me, the setup required that I wear a big jacket, boots, and a cowboy hat, so he wouldn't immediately recognize me. Then I got some help from a neighbor who was out walking her Dalmatian. In full sight of Luke, just outside the property gates, I began giving treats to the Dalmatian and praising him. Luke watched for a moment, excited and interested. Finally he wanted to play so much, he made a beeline across the boundary to join us.

I was ready for him with Level Eight. I turned to face him and I went 'wolf' on him in a dramatic attack mode. I huffed and puffed like I was about to explode, I yelled in a very deep tone of voice, I froze my body and face in a threatening posture, and I thundered and stomped my feet in an Oscar-worthy act. My performance made a big impression on Luke, because he flew back across the boundary, through the gates, and back to familiar territory. He sat and waited while I continued to work with the Dalmatian, until I went back, praised, and released Luke from his sit position.

I don't particularly enjoy using Level Eight with my dogs, and you probably won't either. But I'm willing to do whatever is necessary to save a dog's life, and it often does. Just remember to put everything into it because you may not get a second chance. Luke never crossed that boundary line again. In fact, he refused to do it one time when they *wanted* him to cross, and they had to get the car and drive him through the gate. All of this was accomplished without ever hurting Luke, and to this day he and I are madly in love with each other.

Now you can see why I ask people not to bark commands at their dog. There's a big difference between asking and telling, just as there is a difference between speaking and barking. If you always use a barking tone of voice what will you use in the face of an emergency? Use a regular tone of voice with your dog for the routine comings and goings of life. Then, if you want to exaggerate your joy or your ferocity, your dog will not be confused.

SUMMARY OF COMMON LANGUAGE:

- **Build trust: communicate with your dog in ways he understands.**
- **One word: give one specific word for one specific behavior.**
- **No 'good dog': do not use a generic 'good dog' to praise him.**
- **Silence: your dog needs time to process what he is learning.**
- **Tone of voice: use contrasts in your tone of voice to give your dog clear messages.**

Chapter Seven

Using Treats

When a dog perceives that he is being punished, his fight–flight mechanism is aroused, producing an unpredictable state in his psyche, as well as an inability to learn effectively. Punished dogs show an alarming rise in the stress hormone cortisol. It's a neurotoxic substance that can lead to memory deficits, and chronically elevated cortisol levels can even weaken a dog's immune system. With The Loved Dog method, we don't punish dogs. Instead, we use rewards as motivators, and in this chapter we focus on treats.

Treats are a shortcut to reaching a dog's brain. In other words, the way to a dog's brain is through his stomach! Treats are useful because they are quantifiable. Whenever you want to show your appreciation, you can dispense a larger number of treats, which you can't do with other forms of reward (such as a car ride or a walk). Using treats is the easiest and fastest way to teach a dog what you want him to learn, but you need to keep them special, and therefore highly desired. That means picking and choosing what, where, and when to reward your dog.

The main thing to remember is that you should *never* give treats away for free. They must be earned, or you'll be throwing away one of your most powerful training tools. If a dog gets a treat he hasn't earned, he'll think he doesn't have to do anything to be rewarded. Rewarding your dog for doing nothing will absolutely handicap your ability to coach him.

TREATS AS LEVERAGE

I have seen countless owners giving dogs tasty treats because they gaze up at them with adorable, pleading eyes, acting as if they haven't eaten in a year. How can you possibly deny her a treat when she asks so sweetly? It's easy to fall under the spell of your dog's cuteness, but it won't help her learn good manners. I also see people mixing the yummiest treats with their dog's dry food, to entice the dog to eat. That is giving treats for free! Let your dog's daily food be of the best quality, but somewhat boring. Then you can keep the interesting and enticing treats for games and exercises.

I went to visit Maddie, a pretty yellow cocker spaniel, after her 'mom', Debrah, called for help. It seemed that Maddie was acting out and not responding when Debrah offered her treats to behave. When I walked into the house, one thing was instantly clear. Maddie was a spoiled brat! Dozens of toys were strewn across the floor, but Maddie ignored all of them, and she turned up her nose at treats that any other dog would salivate over. Surprise, surprise. Debrah was genuinely hurt that Maddie ignored the abundance of toys and treats she gave so freely. I explained that she had nothing that Maddie wanted, so Maddie had no interest in listening to her. And like a spoiled child, Maddie was becoming rebellious.

We started by taking away all of Maddie's toys and treats. They would be used for play-training only. Once Maddie realized there was a new sheriff in town, we saw a significant change in her attitude. After the first week, Maddie started to become polite and appreciative. Now she realized that good behavior was the only way she would get any attention, toys, or treats.

The most common mistakes with treats are lavishing them too readily or being too stingy. Your dog needs to earn her pay, which is something dogs love to do. But don't take her efforts for granted, either. Ignoring good behavior is a surefire way to make it go away. The simple truth is that if we reward good gestures, we see more of them.

Specific behaviors call for different amounts of rewards, depending on their level of difficulty. When you're competing with a behavior that the dog would prefer to do in that moment, you need to make your offering irresistible – lots and lots of superyummy treats that are, in your dog's mind, much more exciting than what he is presently doing. There are times when I prefer using toys over treats, but with toys, it's difficult to give them in repetition, which I call 'giving a jackpot'. With treats I can make the lesson more efficient as each reward is gauged to match the level of performance.

For some reason, people have an unconscious tendency to disperse their rewards in sets of threes. You will often see the dog turning away after the third one. Believe me, your dog knows how to count to three. If you become so predictable that he understands that no more surprises are coming his way after the usual three, why should he try harder? Who wants to put out that extra effort when they know that their work will not be appreciated? Stellar behavior merits special appreciation!

GOLD, SILVER, AND BRONZE TREATS

For one of my appearances on the *Today* show, I had to prepare for a demonstration that would use Matt Lauer as the dog – so I did some investigation to find out what kind of treat he liked best! It happened to be chocolate. You need to do the same kind of research for your dog.

In order to figure out your dog's favorites, I suggest you fill some bowls with different treats and put a few of them out at the same time. Take a week or so to find out which your dog likes best, and rank them in order of gold, silver, and bronze. Keep in mind that your dog's taste can change with time, and locations and distractions might play a huge role in her desire for the treat.

Today, there are great companies out there that make nutritious doggy snacks, but don't believe all advertisers blindly. Shiny labels and exciting commercials are not the best indicators of healthy food. I recommend reading the labels on store-bought treats, eliminating anything that contains fat or protein not identified as animal, meat, or chicken byproducts; the preservative BHA; artificial colors and flavors; and sugar or corn syrup. Go for foods that contain ingredients you can recognize, such as whole grains and vegetables. You can add to the list any of your own healthy leftovers; just make sure you wash off any spices or sauces. These could include chicken (no skin or bones), beef, boneless fish, eggs, a little bit of cheese, veggies (no onions), fruits (no grapes or raisins), and brown rice.

Both my dogs love salads with olive oil and lemon, and they both enjoy rice and sugar-free cereal.

When I eat at a restaurant, I sometimes take home a piece of steak or hamburger, which is a great gold treat. Even if you feel weird wrapping a tiny piece of leftover in a napkin, do it anyway. That little square inch of food, a valuable surprise, can become a fantastic and unexpected incentive for your dog. When I eat an omelette for breakfast, I keep two small pieces for my boys to use later in the day. I don't do this every day, but whenever I call Clyde and Duke to come, they are always eager to see what I *might* have for them.

For most coaching situations, I use my special chicken-liver bits, my faithful helpers in winning me leverage over dogs that I have only just met. In all my years of training, not one dog has ever refused them, so they are my sure thing. But even with a sure thing, it's important that your dog doesn't get bored with the food, or he'll be much more interested in chasing a bird across the yard than learning how to 'back off'.

When the stakes are the highest, such as trying to train a dog in the middle of distractions, or when she's learning something new for the first time, I use gold treats. A gold treat may be my special liver treat or a rotisserie chicken that I deboned and cut into small pieces. I also use beef or a hot dog – always lean – when I'm competing for a dog's attention around other dogs, such as at the doggy park or on a busy sidewalk.

Silver treats, such as a piece of hard-boiled egg or a dog biscuit, are for medium-level circumstances, like repeating already learned behaviors. If you're doing a regular job, you get paid a regular salary. No bonuses.

Bronze treats, like veggies dipped in olive oil, fruit, or dog food tossed in a plastic bag to mix with the leftover powder from gold liver treats, are used when the training challenges are at a minimum. They are also good for keeping your dog at a healthy weight, which is important to remember when using lots of treats in your training.

Remember that using the same dog treat again and again will become boring, no matter how delectable it was for her at first, so mix it up. Just vary your dog's treats every so often and make sure you keep them unpredictable.

STORAGE HAND

You need to keep treats readily available so you can reward your dog immediately – no longer than three seconds. That's all the time you have to make sure he's clear about what he just did correctly. If more time goes by, he'll be on to the next thing, and think he's getting a reward for nothing. If you stop to finish a conversation and then reward your dog, you missed your chance. This is also true for negative feedback, because if you wait more than three seconds, he'll have no idea why you're disappointed or upset with him. If you fail to act immediately, the opportunity will be gone. Communicating with him in a way that he can understand requires lightning-speed feedback.

When I do my coaching, I always keep two or three different treats ready in my storage hand – the hand I'm not using for hand signals – broken into small pieces. I also put treats in containers and keep them around the house, wherever I spend time with my dogs. Keep containers of treats wherever you and your dog hang out – the living room, kitchen, or bedroom, and always by the door, where dogs get the most excited. Wherever and whenever you actively coach your dog, always hold enough treats to instantly reward your dog when he shows you how brilliant he is.

RANDOM REWARDS: THE LAS VEGAS METHOD

There is a myth that if you train your dog with treats, the dog will stop listening to you once you stop using them. People ask me all the time if they'll always have to give their dogs treats to get them to listen. The answer is no. In the teaching phase, I have to use something. But once they understand what I want and give it to me, I immediately switch to what I call 'random rewards' or the 'Las Vegas method'.

I'm not talking about weaning the dog of rewards altogether. A dog just needs to know that there is always the possibility of a jackpot. Just like playing the blackjack table at a casino, you only need to win once to think you might win again – and keep playing! If a dog knows that a gold treat *might* be there, that becomes the jackpot that inspires him to keep focusing and behaving as you ask him to.

In order to decide how to work with random rewards, let me explain the two phases of coaching your dog: the teaching phase and the everyday phase. In the teaching phase, I reward the dog for every success, large or small. During teaching, all successes are important. I use a tasty gold treat and a happy tone of voice, and I'm anxious to find any opportunity to praise him. I focus my attention on anything good that the dog is doing right. When he does something faster or better, I give him a quick series of gold treats – five, six, or seven in succession, using my most excited tone of voice. That's the jackpot – he just pulled the handle on a slot machine, heard that jingle, and it paid out a big reward that never seemed to stop.

The next step is very important. When he begins to understand what I want, I switch from rewarding every success to rewarding the dog's behavior randomly, but still giving him jackpots for exceptional behavior. If I give the dog a treat every time he does what I ask of him, his behavior will start to get worse, not better, because there will be no excitement or surprises. Psychologists call this 'random rewards', and gambling is the perfect example. People have a hard time quitting after they win even once. The idea is that we tend to take for granted whatever is certain in our lives. If I get rid of treats altogether, or if I always give the dog a treat for each and every thing, he'll lose the fever. Renowned behavioral psychologist B.F. Skinner discovered that when you give animals random rewards, they are more eager to learn than when you regularly give rewards. Random rewards create appreciation, and they are never taken for granted.

The right time to switch from the teaching phase (giving treats every time) to the everyday phase (random rewards) is when the dog consistently understands what you want him to do. Some dogs will grasp the idea in a couple of minutes, and others may take a few days. It doesn't matter how long it takes. All that matters is that you switch as soon as your dog understands.

When you give your dog random treats for good behavior, he'll think, *Maybe this time I'll win the jackpot!* and will continue to do what you ask of him. The difference between gambling in Las Vegas and training your dog, however, is that in Vegas, the house always wins. With The Loved Dog coaching method, everyone wins!

RELEVANCE TRAINING

Each element in your dog's daily routine should involve relevance training. For example, anything that he wants, he needs to sit for. Before your dog gets his breakfast, ask him to sit. Before you play a game with him, ask him to sit. Before you put on his leash, open the door for a ride in the car, or cuddle with him, ask him to sit. This way, coaching is not an isolated exercise that you do several times a day for a few minutes. Rather, it becomes a part of the very fabric of his life.

My Dukey sheds a lot. When I first got him, he did not like being brushed, but he needed it, so I used relevance training. Since he loved to go on daily hikes, I asked him to let me give him one really short brush before I opened the door. Over time, I brushed him for longer periods. Now, the brush excites him, because he knows it means a *hike* is coming! Through relevance training, you can eliminate giving your dog treats for everything. Once he sees the relevance of doing something to get what he wants, he'll get the point. Remember to give him a jackpot once in a while to keep him enthusiastic about training, especially if he does an excellent job.

Make a list of all the things your dog likes and that get him excited. It might include a walk, a massage, a kiss, a treat, a chew toy, a tennis-ball toss, or an invitation to join you on the couch or the bed when you get home. Then figure out which exercise goes best with the things on your list. I teach my dogs to earn and appreciate every reward! Appreciation is something that must be taught, learned, practiced, and integrated into the everyday life of our dogs.

SUMMARY OF USING TREATS:

- **Treats as leverage: treats are both payment and motivation for your dog to learn new behaviors.**
- **Gold, silver, bronze: break down treats into three categories.**
- **Immediate feedback: give a treat within three seconds of the behavior.**
- **Las Vegas method: use random rewards to keep your dog focused and interested.**

Chapter Eight

Socialization

Without socialization and learning proper manners, a dog's outlook on life will be determined more by her wolf aspect than her toddler aspect. Living in the bustle of human society, meeting other dogs while on a leash, and being approached and touched by strangers – it's all unnatural for a dog. Desensitizing her to these experiences is where her domestication actually begins.

It's best to start socialization in puppyhood, but contrary to popular belief, it is not just for puppies. Socializing your dog is a lifelong process through which any dog can be prepared for every situation that may arise. We cannot look to society to learn how to deal with our dogs; we must teach our dogs how to deal with society. Throughout the years, I've seen many common behavioral problems in dogs that could have been prevented with proper socialization.

We have already established that the force behind your dog's behavior is what she links to pain and pleasure. Therefore the goal is to teach your dog to associate everything in her daily life with pleasure. That includes different types of people, such as tall, short, bearded, uniformed, male, and female. It also includes different types of loud noises, like car horns, the vacuum cleaner, clanging garbage cans, or a backfiring car. Dogs also should be socialized to cope with as many kinds of children as possible, from infants to teenagers. Kids can be overwhelming for *us* when they are rambunctious or excited, so imagine how overwhelming they are to a dog. Then add the fact that dogs are often the object of children's curiosity and demonstrative affection.

Most people think of socialization in terms of taking the dog outside and introducing her to people, kids, and other dogs, but that is far from the truth. Just because your dog is exposed to different life experiences does not mean that she's ready to tackle any challenge. Exposure does not equal socialization. In order to properly socialize your dog, you must connect a pleasurable association in the dog's mind to those experiences.

If your dog demonstrates any signs of fear, they must be addressed immediately. A fearful dog will freeze, fight, or try to take flight. Fear is often accompanied by a growl, which most people translate as a bad thing that needs to be stopped. People think that stopping the growl will stop the escalation to a bite. Wrong! Stopping a growl simply removes that useful alarm signal. Now the already agitated dog gets even more stressed, worrying that his owner may be upset with him on top of what is already wrong.

So what should you do? The moment you hear your dog growling, get her out of the situation and assess what she is afraid of. Many people are offended when they hear their dog growling at them, but it is never personal. Getting incensed, yelling, or hitting a growling dog will only lead to an escalation. She'll feel that people are even less trustworthy than she thought, and she will become even more dedicated to protecting herself. But now she'll do it without a warning growl, so watch out!

Dogs growl for reasons other than fear, but it is usually the main cause. We all know that the best defense is a good offense, and if the growl causes a particular threat to back off, then in the dog's mind, it worked. Since a rewarded behavior is likely to increase, your dog could become an expert growler once she realizes that her tactic removed the threat. Unfortunately growling sometimes can escalate to full-blown aggression.

Avoiding this escalation of fear is what socialization is all about. Turn your dog into a canine ambassador, and remember that if she is afraid of something, don't punish your dog or try to correct her. Instead desensitize her by using baby steps, and change her painful associations into pleasurable ones. Be sure to use socialization exercises inside the house in different rooms, outside on the street, and during the daytime and at night. Hardly anyone trains their dog at night, but why not? The dog needs to be socialized when it's dark outside just as much as when it is light. Sudden interruptions can occur at any moment, so socialization applies at all times of day and night, inside and outside, whether your dog is walking, running, or lying down.

DESENSITIZATION

Start by having a gold treat ready for each time your dog sees a new dog, person, or even the scary garbage truck. It may sound counterintuitive to give your dog a treat when she's afraid of someone or something. You may think that giving a treat is rewarding her fear. However, you're using a method commonly accepted by psychologists: minimizing the fear element as much as possible while increasing the pleasure element associated with that particular fear. To think a dog should just get over it is as useless as asking somebody to get over his fear of flying or heights. If your dog is afraid of something, gradually introduce her to it. Most fears arise when a dog feels out of control, and socializing the dog minimizes these feelings.

A couple of weeks ago, I got a call from my friend Jamie, who needed help with her thirteen-year-old Lab mix, Daisy. Daisy was afraid of Melissa, the masseuse, who visited the house weekly. Luckily, Melissa was willing to help out, so we asked her to come to the home as many times as possible to feed Daisy, take her on five-minute walks, and give her treats and toys. We also put one of Daisy's beds in Melissa's car, so that it would pick up her scent. The next time she came to give Jamie a massage, Melissa carried in the 'scary' massage table, and we plied Daisy with treats. Before long she was happy to see Melissa coming, thinking that the massage table had turned into a treat dispenser. Whether it's a neighbor, the UPS person, or the pool guy, the same desensitization techniques apply.

There are many sounds and experiences that a dog might fear, not only outside the house, but inside as well. There is the roar of the vacuum cleaner, dropping a book on the floor, or shaking out a plastic trash bag. Always have a treat ready, inside or out, so when your dog hears an ear-shattering sound or meets someone new on the street, she'll be rewarded. It means a treat and a pleasurable experience for her.

Once she has transitioned from an association of fear to an

association of pleasure, your dog's encounter with anything unusual, loud, or threatening will only make her think, *Whatever . . . Now, where's my treat?*

PHOBIAS

I worked with a beautiful golden retriever puppy named Snickers, and when I arrived at her home, I noticed right away that she was cowering in a submissive body posture. When she came toward me with that sweet golden puppy face, she was practically dragging herself across the floor, trying to stay as low as possible. I discovered that the simple act of opening a plastic garbage bag, a daily activity in most homes, absolutely terrified her. In fact, when I crinkled the bag several feet away from her, she submissively urinated. I knew immediately that the idea of teaching Snickers new words and play-training exercises was too advanced. First and foremost I needed to help her overcome her phobias.

As a puppy, Snickers had been kept in a crate all day with only a few bathroom breaks. This was not done out of meanness or cruelty, although ultimately it resulted in a terrified, skittish dog. The couple that owned her had been told this was the correct way to housebreak a dog, and Snickers bore the brunt of their misinformation. Furthermore her owners had gotten her from a pet store, which meant that she was the product of one of the most inhumane, abusive, unhealthy, and debilitating practices of all: a puppy mill.

Now that I understood Snickers a little better and took into consideration her puppy-mill background, I began to work with her fears, starting with garbage bags. Remember that desensitization is about decreasing painful associations and increasing pleasurable ones, which is what I set out to accomplish with Snickers.

I shook out a bag several times, and each time Snickers ran away and cowered in the corner. She didn't want anything to do with that bag – or me, the person who was shaking it. Unexpectedly I offered her a gold treat. *Hm, what should I do?* she thought. I did not move toward her. I just kept the treat in my hand, luring her toward me. She slunk forward in terror, keeping her tush as close as possible to the floor. She made it to me, took the treat, and hurried away. She was terrified that she was making some kind of lethal mistake. But I had her attention, and the healing process had begun. While I was giving her the treats, I did not move the garbage bag at all.

Some of the most common fear indicators in a dog are panting (the doggy equivalent of sweating), yawning a lot, licking lips obsessively, frequent urination, and unexplained diarrhoea. Snickers's owners had no idea how much fear their dog was displaying. My clients often ask, 'Isn't this kind of fear normal? Won't she grow out of it?'

'No,' I tell them. 'It isn't. Not only will she *not* grow out of it on her own, it will only get worse unless you socialize her. We have to desensitize her and help her realize that life is good.'

Each time I stood up and crinkled the bag for Snickers, she rushed away. I began to place treats closer and closer to the dreaded garbage bag on the floor, saying, 'take it' each time she gobbled up a treat. Using a calm, happy voice and what I call 'safe words', I was building a reward history in her mind. Little by little I helped her to associate pleasure with the words 'take it', Snickers happily ate each treat I gave her, and all the while I repeated the safe words 'take it'.

Then I put a treat directly on the bag. She refused it – an indication that I needed to make the challenge easier for her to overcome. I placed the treats a few feet away from the bag, moving them closer and closer each time. Finally Snickers worked up the courage to take a treat from the corner of the bag, which resulted in a huge jackpot that I placed on the corner of the bag. I 'made a party', repeating 'take it, take it, take it!' In a few more minutes Snickers was eating treats right off the bag, even when I moved the bag around!

In less than ten minutes Snickers understood that whenever she heard the words 'take it', she could expect a pleasurable experience to follow. Now, 'take it' had become her safe words. I avoided saying 'good dog' and I also avoided saying 'it's okay' or 'it's all right' when she backed away, so she wouldn't think I was praising her fearful behavior. I know it's tough not to feel sorry for your dog, and you want to comfort her in stressful situations like this, but it won't help her to overcome her fear. In fact, babying your dog while she's scared only perpetuates her phobia.

For the next few days, I asked Snickers's owners to open and close doors loudly, drop heavy objects on the floor, and run the vacuum cleaner. They had a piano, and I suggested that they should bang the piano keys loudly, all the while giving her gold treats. Now it's up to her owners to continue the training and consistently build on it, until all of Snickers's fears are things of the past.

TOUCH AND FOOD PROOFING

It's important to teach dogs not to be worried, scared, or aggressive when strangers invade their personal space. As preparation for the real world, dogs should be conditioned to be touched all over their bodies. Vets, groomers, and children will want or need to touch your dog, and he needs to be okay with this.

I worked with a handsome Bulldog named Winston who needed to learn that touching was a good thing. Each time I gave him a treat,

I pulled very gently on his ear, touched him between his toenails, or I gave a playful tug on his stubby tail. He easily learned not only to accept all kinds of touching, but even to enjoy it. Keep in mind that in a wolf pack, touching occurs according to the leader's wishes. A lower-ranking animal is not allowed to touch the leader without an invitation. Only the leader can initiate touching, and by doing this you are communicating that you are in charge.

I spent some time in a canine rescue facility in London that helped owners solve their dogs' behavioral problems instead of simply dumping their poorly trained dogs into shelters. When we asked them to do something as simple as brush their dogs in front of us, the truth came out. Most people admitted that they were not really moving to an apartment that would not take pets, as they had initially explained. The dogs were impossible to control and couldn't even be brushed without a fight. The owners were at their wits' end, and felt that they had failed.

The trainer at the facility demonstrated how to get the dogs on a table that was positioned next to a doorknob. He put a leash on a dog, secured it to the knob, and fed treats from one hand while he brushed with the other. At first the grooming session lasted only a few seconds. Over time, though, the dogs began to realize that getting brushed was the only way they would get the gold treats. After a few sessions they didn't mind being groomed, and even started enjoying it. Happily, these dogs remained in their original homes.

Eating is another activity in which a dog needs to be desensitized. I don't know how many times I've heard someone say, 'Don't bother the dog when he's eating.' A dog that is not socialized is not used to having company during feeding time, and he may become alarmed and feel the need to aggressively protect his food. The dog needs to accept and eventually *like* being disturbed while eating. We do this by showing him that when someone approaches his food bowl, it's a great, fantastic experience.

I taught a chocolate Lab mix named CJ not only to tolerate me interrupting his meal, but to look forward to being disturbed. As with many of The Loved Dog methods, I coached the dog in stages. I started the bronze phase by repeatedly walking toward CJ while he was eating, then dropping a bronze treat into his food bowl and petting his back. This got him used to the idea that an approach from someone *could* mean a treat. I made sure that CJ was not freezing or tensing up, and he was pretty relaxed when I came close, hovered over him, and dropped the treats into his bowl. He was probably thinking, *It's raining treats! Hallelujah!*

Once the first phase was successfully completed, I moved on to the silver phase.

Since I already knew CJ did not mind my hovering, I brought my hand closer to the dog. Then I put my hand into his bowl while he was eating, and I gave him a silver treat. This showed him that my interruption meant he would get something yummy. It was crucial for him to trust that I was not taking his food away. Rather, I was improving it. We practiced this a lot, and when he was excited to see my hand in his bowl, he was ready for the gold phase.

While CJ was eating, I interrupted him by picking up his bowl and dropping in an amazing gold treat: a big piece of chicken. I asked him to sit, and I put his bowl back down, which was now much more delicious than before. He quickly understood that each time I took his food away, I was improving it and he would get it right back. Now he *wanted* me to interrupt him! When you socialize your dog in this way, many bites can be prevented. You are teaching him that he has no need to be possessive, no matter how much he treasures his food. Ultimately he is learning that people may disturb his meal, but it's okay because something good will happen.

NIPPING

Nipping is acceptable behavior in a wolf pack, but it is very bad manners in our human society. My client Erin got cuts and scratches all over her hands from her puppy, Honey, a Goldendoodle. She wasn't mean or angry. She was just demonstrating a wolf's playful behavior when Erin rewarded her for good behavior with treats or affection. But when a dog's mouth is next to me, I want to feel either kisses or nothing. Teeth are not an option.

I needed to teach Honey to kiss hands, not bite them, so I pulled out a jar of peanut butter. You can use any healthy food that can be smeared on your hand. Some dogs like peanut butter, and others like honey or liver spread. I smeared a thin layer of peanut butter on the back of my hand and put it out for Honey to lick. As soon as she started licking the good-tasting treat, I repeatedly said, 'kisses, kisses, kisses', with a soft and appreciative voice. 'Good kisses,' I reinforced, as Honey savored the peanut butter. I was teaching her that licking my hand, or kisses, was something I wanted her to do, without correcting her nipping behavior.

Even though Honey liked the peanut butter and understood that I wanted her to kiss me, she continued to nip from time to time, so the next thing I used was negative feedback. When Honey nipped my hand, I let out a high-pitched 'Ouch!' to let her know that I was not happy. I made my voice sound similar to a dog's yelp – the kind of noise he would make if his playmate nipped him too hard. After Honey nipped my hand and I yelled 'Ouch!' she withdrew. I gave her my hand again for a second chance, and she kissed me – she got it! From there I moved to getting kisses on my bare feet, and ultimately I got her to kiss my face.

It is important to teach your dog 'bite inhibition', so he learns to do everything with a soft mouth. When I put a treat in front of a dog's mouth and I can feel his teeth when he tries to take it, I scream out 'Ouch!' Even if it doesn't hurt, it could have. I refuse to continue playing until he licks me or gives me his gums or his tongue as he's reaching for the treat. I can see the dog's face, as if he's saying, *Okay, I'll try to do it softer. I didn't realize that people were so delicate.*

I recently went to train a cute little boxer puppy, who had started lunging and biting at people. It turns out that the father of the house had started wrestling with him. He admitted that he'd approached the dog and touched his head, egging him on with growling sounds, as if trying to talk doggy to him. The next day the boxer couldn't wait to play his favorite game, and each time his owner obliged, the dog bit harder than before. In a very short period of time, the boxer got so good at it that he began looking for additional playmates by lunging at anyone he met, a behavior that appeared aggressive.

The boxer also added the wolf game 'catch me if you can' (running around like a crazy dog). He would rush to his 'dad', bite him on the arm, and run away again, thinking he was playing the greatest game in the world. After all, it was in his nature to practice his hunting skills by playing like this. One afternoon he decided to play with one of the kids who didn't think it was any fun at all. When the young boy reprimanded him, the dog got very confused. *My dad loves this game. What's wrong with you?* he must have been thinking. *Will you change your mind? Let me try again.* He dashed in for another bite and starting running around like a lunatic again.

Even though the puppy was small enough not to hurt the father of the house, what would happen if he played that game and bit a neighbor or a child? We have to teach our dogs that under certain controlled circumstances, they can wrestle with their own species, whose thick skin is covered with fur and won't sustain injury. We want them to understand that we are delicate creatures and that the tiniest nip with a tooth on our skin or even on our jeans can cause us severe pain.

If you yell 'Ouch!' and your dog still doesn't stop, get up immediately, leave the room, and close the door in your dog's face. Nobody likes to lose his playmate. After a couple of minutes, come back to the dog and play with him again. It may take time, but dogs catch on. The separation and isolation will make your dog think twice. Another way to stop a dog from nipping is to play tug-of-war, a game I describe in chapter 11. I am absolutely against hitting your dog under the chin or holding his mouth shut, like many trainers recommend. I want a dog to associate people's hands with pleasure and not threats. Using violent means to solve a nipping problem will only make your dog act defensively, or even aggressively, when he sees you or anyone else coming toward him.

GREETINGS

We greet our friends and families differently than dogs do. From an early age human beings learn to look directly into someone else's eyes and stretch out a hand of welcome, which is considered an appropriate gesture of sincerity and politeness. But this is not the case for dogs. A dog considers a direct look in the eye as a threat or confrontation. When you're working on socialization, never look a fearful or shy dog directly in the eye. If a fearful dog can't remove himself from a situation, direct eye contact may cause him to display aggression. This is not dominant, territorial, or possessive aggression, but nervous aggression, which is an attempt to cause the threatening person or animal to retreat and increase the distance between them.

Notice whether his body is stiff and rigid or relaxed and happy. Body stiffness is a signal that he's *very* unhappy, almost ready to attack, even if his tail is wagging. There's a scent gland in your dog's anal area, and when he's afraid, he'll try to cover it by tucking his tail under. This is why when some dogs are scared, they'll perform that silly-looking walk-and-sit-at-the-same-time maneuver. An arrogant dog walks with a stiff tail held high. A truly happy-go-lucky dog will hold his tail at midheight, wagging it loosely and smiling. (Dogs really do smile, as well as use their mouths to show dislike or fear.)

Although it will be counterintuitive for you, I suggest you face sideways. Put out your hand with a treat and *let him come to you.* If he is too fearful to accept your hand, keep your eyes averted from his face. Without making eye contact, drop the treat close to him on the floor. Little by little you can bring the treat closer to you, until eventually he'll learn to take it from your hand. Use safe words like 'take it' when he does.

Next, try teaching him the exercise 'watch me', when he gets rewarded for looking in your eyes. Simply hold a silver treat between your eyes with your hand. When the dog looks at you, say, 'watch me', and reward him. Little by little he will realize, *Although humans are rude to look directly at my eyes, I actually like it. It's not scary at all. What was I thinking?* This is a great skill to develop for times when you really need to get your dog's attention.

What do you do when a child rushes over to your dog to kiss him on the face? Children are drawn to hugging furry animals. However, dogs are not always so anxious to be hugged, particularly when the love assault comes out of left field. A little girl's first impulse might be to greet the dog with hands stretched out, running toward him, the way she was taught to greet other people. But all that in-your-face business can be too much for the dog to take, and he may snap or even, God forbid, bite. She may look adorable to you, but unless the dog has been successfully socialized, he will see her as a threat or a potential attacker. His options are freeze, fight, or flight. Since he will probably be on a leash, flight is not possible, and freezing will not stop her from coming over to pet him. All that's left for him to do is fight, unless you can communicate to him that he's safe. If he's trying to get away, don't force him to interact with the cute little girl, or whoever it may be.

I give the dog a jackpot as soon as he meets a new person, changing the dog's freaked-out reaction to one of pleasure. Suddenly getting hugged and petted isn't so bad, since it triggers a jackpot of gold treats. I want to increase his pleasure when a stranger shows up, so he will be eagerly waiting for the next stranger.

When you're approaching another dog, stop moving toward him long before you see signs of agitation. Always keep a good distance and don't look *directly* into his eyes. If he wants petting, he'll come to you, particularly if you're bending down with your body at a forty-five-degree angle to him. So wait and assess the situation. Let him sniff your hand and decide whether he wants a new friend or not. Just to be safe; always ask an owner if his dog is friendly. If you're not sure, forego greeting a strange dog. You just never know how he's feeling.

MEETING NEW DOGS

Vets do not all agree on the right time for a puppy to meet new dogs. Some believe a puppy should be kept indoors with no contact with other dogs or the outside world until he is finished with all of his puppy shots, which is usually by the time he's sixteen weeks old. Other vets feel that he should be socialized earlier. Puppies are most impressionable during these first sixteen weeks, which is a small window of time to introduce new things. The way I look at it, there are way more dogs put to sleep from lack of proper socialization than dogs who die from catching a deadly disease. I believe that we should keep our puppies at home until they have completed their second set of vaccinations. After that, I encourage people to invite a variety of vaccinated, friendly dogs to their home to play with the puppy. After the third set of vaccinations, I encourage people to take the pup out on walks, but not to let him go near faeces or to come in contact with unfamiliar dogs until he gets his fourth, and final, set of vaccinations.

Sometimes, if you're lucky, an older dog will help you with the training. Have you ever seen a pesky puppy jumping at an older dog who corrects him with a threatening growl? We react as if the animal who gave the correction is wrong, but he is *right*, and we need to leave them both alone. Dogs will teach one another what is acceptable and what is not. Just like with kids, not every pet wants to play the same game at the same time. If your puppy nips at an older dog and is corrected with a growl, you should thank that older dog. He just did you a great service by teaching the rules of the game.

The older dog is preventing your puppy from using aggression and inappropriate pushiness. Let the old guard show your puppy how to play politely. He can then enjoy the company of other pets while you can enjoy the company of their owners.

SUMMARY OF SOCIALIZATION:

- **Not just for puppies: socialization is a lifelong process.**
- **Address fear: if you notice that your dog has unreasonable fears, help him overcome them immediately.**
- **Touch proofing: create an association of pleasure when it comes to being touched by children, vets, and groomers.**
- **Disturb when eating: teach your dog to look forward to mealtime disturbances.**
- **Greetings: show your dog that meeting strangers is a pleasurable experience.**
- **Meeting new dogs: after your dog has been properly vaccinated, set up a puppy play-date.**

Chapter Nine

The Power of Sit

I consider 'sit' the foundation of my work. It's the trunk of the tree, and all other behaviors are like branches growing outward from there. Many trainers think the first training step is to put a leash or choke chain around your dog's neck, take him outside, and teach him to heel. But if he doesn't even know how to sit, how will you get his attention in the midst of a million smells, sounds, and motions all distracting him, making you the least interesting influence in his environment? You must establish a connection with your dog, and the first step is teaching him how to sit. When your dog feels secure with you as his leader, he'll be happy to stay close and follow your lead as you introduce new challenges. Teaching your dog to sit without force or a leash is easy since it is something that he does naturally.

I recently worked with a two-month-old white Doberman puppy

named Flash. I guess his owner, Eddie, was expecting a burly male trainer with a bag of torture devices and a rough demeanor. What he got instead was a petite woman with a bag of treats and a penchant for 'making a party' every time his dog did something right.

'Does he know how to sit?' I asked, ignoring Eddie's obvious disappointment in meeting me.

Eddie sniffed. 'Of course. There's no need to start there. Flash sits in every corner of the house all day long. I need him to learn to heel! When I take him for a walk, he pulls me down the street. How can I stop him?'

Since Eddie and his wife had three children, I crouched on the floor at the height of the youngest child and asked the dog to sit. He looked at me like he'd never heard the word before, since he thought that 'sit' only happened when someone stood to face him. It took him twenty minutes, but when he finally got it, I went to town 'making a party', feeding him a luscious jackpot and using my voice as if it were confetti.

'Flash didn't really know how to sit, did he?' asked Eddie sheepishly.

'No, he didn't,' I said. 'Do you see why I didn't want to take him outside to heel first? He needs the foundation of "sit". Then we can teach him anything.'

I'm sure you can appreciate how amusing and satisfying it was to see Eddie, a grown man in a suit and tie, lying on the kitchen floor beside me, 'making a party' for Flash!

I remember when I first picked up Duke from the vet and took him to Covenant House. Although he was excited to perform a variety of exercises, when we asked him to sit, he ignored us. I realized that his previously broken legs made sitting on request uncomfortable, so I showed the teens in the program how to do what I call 'passive training'. It requires no effort – only awareness and treats.

Each time Duke sat on his own accord, we praised him with a treat and enthusiastically sang out the word 'sit!' as if he had just climbed Mount Everest. Kids are wonderfully open about using their voices melodically, and so after a few 'singing lessons' with these talented young adults, with no force or pressure, Duke learned to sit when he was asked. What's more, sitting became his favorite thing to do, because he loved the big fuss they made.

Many years ago I went to observe a class that was held at a local pet store. One of the dogs, a German shepherd, refused to sit. While each owner was practicing with his own dog in the class area, the trainer decided to help the shepherd get over his 'stubbornness'. But no amount of choking, jerking his neck, or verbally commanding and shaming would convince the dog to sit. The trainer did every abusive thing she could think of, including poking and pushing with all her weight on the dog's rump. He still refused to sit, and after a stressful and violent session, he ended up biting the trainer on her hand. Fearing for the dog's life, I got his owner's number and called a few days later.

'How is your dog doing?' I asked.

'He's with the vet,' said the owner.

'Why? Are you going to give him up?' I asked.

'No,' the owner informed me. 'He had an abscess on his rear end and he's getting treatment for it.' The poor dog was suffering with a painful abscess, but he had no way to tell anyone. When he just couldn't take it anymore, he finally reverted to biting as self-preservation.

I've seen so many people turn the other way when their dog is giving them a clear message that something is wrong. I once watched a terribly disturbing segment on CNN that profiled a woman who sat by while a so-called dog trainer beat her seven-month-old German shepherd puppy to death because he was too rambunctious to obey. The owner said that the trainer gave her the following heads-up: 'Your puppy may cry, try to escape, or pee when I'm training him, but that's all normal.'

Give me a break! There is *nothing* normal about those behaviors. If you are doing everything right and your dog refuses to do something, don't beat him up. If he refuses to do something and there is no apparent reason why, I recommend taking your dog to the vet for a check-up. If he gets a clean bill of health, then take him home and start the training all over again. But you should never revert to pushing, shoving, or pressing on your dog to get him to sit. When you force your dog into a position, the stress involved will inhibit his ability to learn and figure it out for himself, not to mention the mental and physical abuse you are inflicting upon him.

THE MAGNET

The easiest way to teach a dog to sit is a method I call 'the magnet'. Hold a treat in your hand, covering it with your fingers. Make sure that nothing sticks out, so the dog can't steal it, and move it back over the dog's head, toward his tail. The dog will follow it with his head back as if your hand is a magnet. Because of the way a dog is built, lifting the treat from his nose up slightly toward the top of his head will get him to sit. In this kind and gentle way, I can instantly move 99.9 percent of dogs into the sit position without a word or a touch.

Holding the treat too high is a common mistake. One pointer's owner kept holding his hand so high above his dog's head that the dog had to jump to get to the treat. I found out pretty quickly that he was afraid that his dog would nip at him. As a result, every time the dog reached for the treat with his mouth, the owner jerked his hand back. The dog must have been thinking, *This jumping business is fun! It's so easy to snatch that treat from you.* As you can imagine, his jumping skills improved while his sitting skills all but disappeared. I needed to stop the work and focus on helping the owner get rid of his fears of being nipped. Then we had to change the hand signal for sit, since the dog thought that meant to jump on his owner. It took only one day to resolve this issue, and the pointer was well on his way to becoming a well-mannered dog. Remember that if you're doing the magnet and your dog is jumping to get the treat, you're probably holding your hand too high. If he's backing away instead of sitting, practice the sit against a wall, so he has nowhere to go.

Be encouraging, patient, and sweet throughout the learning process, and remember to keep your fingers securely wrapped around the treat. If even a corner is showing and available to your dog, he will try reaching up and grabbing it from underneath. Never underestimate his intelligence to figure out how to get what he wants. As soon as your dog sits, give him the treat and introduce him to the word 'sit' in a happy singsong voice, repeating the word over and over: 'sit, sit, sit'.

Like the example above, first I teach the pattern, and only afterwards do I give it a name. Let your voice show your dog that you approve and are impressed with his genius behavior. Soon you'll see your dog thinking, *Boy, am I a good people trainer! Each time I sit, I get a goody, so I'll do it as much as I can!* Now your dog is asking you to train him! Kids as young as two or three can practice this. It will teach them to be calm around the dog, and it can help kids who are afraid of dogs to overcome their fears.

THE THREE LEVELS OF SIT

The lesson is not over once your dog has learned that 'sit' means 'putting his tush on the ground'. As Eddie discovered with his Doberman, Flash, many dogs only respond to 'sit' when their owner is standing up. Years ago, at a seminar for dog trainers, the instructor asked the participants to stand up and ask their dog to sit. The trainers scoffed at how ridiculous and elementary that request was. The feeling around the room was 'Come on, we're professional dog trainers!'

Undeterred the instructor then asked everybody to lie down on the floor and tell their dog to sit. Can I tell you how many confused dogs were running around, trying to figure out what their owner was asking of them? The dogs thought that 'sit' meant that when the person in front of them stood, they would drop their tushes to the ground. But if the owners didn't give them the usual physical cue, the dogs were confused. Even the most savvy of professional trainers can fall into the trap of thinking that he is teaching the dog one thing when in fact he is teaching him something else. This is why I advise you to generalize each exercise as much as possible. Your dog will learn to recognize the signal that you are giving him rather than just the context in which you usually give it.

If you wonder why on Earth you should teach your dog to sit while you're lying down, think about different scenarios in which you might be on the floor – playing with your children, practicing your yoga routine, or hanging out on the lawn with friends. Wouldn't it be nice to enjoy what you're doing and know that you still have control as the leader? You'll never have to worry that your dog will climb all over you.

TEACH SIT FROM A STANDING POSITION

Since we need to train the dog to sit from all levels, let's start with the easiest – the standing position. I always start with the easiest first, so the dog's success will bolster his confidence for the tougher challenges to come. Begin teaching sit by using a hand signal, which is easier for the dog to understand than your voice. Only later will you teach the verbal cue. Utilizing the magnet, raise your hand above the dog's head with a treat. Practice this until your dog starts to sit correctly on a regular basis. Then repeat just using a hand signal – the exact same motion, open palm facing up, but without a treat. To teach the verbal cue, say the name of the exercise, sit, but only after the dog shows you the behavior, and then offer praise.

Once both cues are learned, common sense will indicate when to use a hand signal and when to use your voice. If your dog is running away from you, and you want her to stop, how will she see you giving her a hand signal? She needs to hear you.

TEACH SIT FROM A SITTING OR CROUCHING POSITION

The second level of sit is when the owner is sitting – and I don't mean on the couch! I worked with Honey, Erin's Goldendoodle, on learning how to sit, and when I crouched to the floor at sitting level (the level of a standing child), she immediately jumped on me. She was happy to see me lowering my body because in her wolf language that meant 'Let's play!' I kept turning away while not looking at her, ignoring the behavior I didn't like. I did the magnet with her until Honey began to get the point. She learned quickly to sit when we were at the second level, but at the third level, when I lay down on the rug, it was back to square one.

TEACH SIT FROM A LYING DOWN POSITION

With Honey I immediately reverted back to the magnet, right from my position lying down on the floor, raising my arm, holding a yummy treat as high as I could, an inch above her nose. Honey immediately recognized the pattern that she had learned when I was standing and crouching. She followed my hand and in no time, she had learned to sit while I was lying down – without me having to ask her. Now that's what I call manners!

Think about how many times you're sitting or lying on the floor, and your dog begins to climb all over you. I got a call a while back from singer/songwriter Olivia Newton-John to help her with her beautiful Irish setter, Jack. She liked to practice her yoga exercises on the floor, and Jack saw this as his opportunity to climb all over her, loving and kissing her. That may sound cute, but, in fact, it became annoying and even potentially dangerous.

I taught Olivia how to do the magnet with Jack when she was lying down, and instead of jumping on top of her, he sat. When she reinforced the word, 'sit', Jack started to get the picture, especially since Olivia kept turning her back on him when he didn't. When she was finally able to say 'Good sit!' and give him a treat, he got it and sat every time. Now he can be with her when she's exercising instead of being exiled to another room. When your dog has manners, he can share more of your life with you.

I worked with a Rhodesian Ridgeback mix named Moby, whose owners, Matt and Abby, had won him in a game show on Animal Planet called *Who Gets the Dog?* in which I was the behaviorist judge. After a two-week honeymoon period, they discovered that Moby was a handful! This is not unusual. It's easy to get along with anyone (including a new dog) at first, when they're trying to make a good impression. Matt and Abby had no idea that Moby was gathering data about his new family that would help him manipulate his people and his surroundings. (Did I mention CIA agent when I listed self-appointed jobs for your dog?)

Soon after Moby moved in, Matt, an army journalist, was deployed to Iraq, leaving Abby all alone with this supersmart and superenergetic dog. When I met with them two years later, Abby was close to giving birth to their first child, and it was urgent that Moby learn how to behave gently and politely. We used the blanket on which they intended to put the baby when she was on the floor. Since Moby was so hyper and unruly, it started out as a big challenge. But using my magnet method and positive reinforcement, Moby happily learned to sit when Abby was lying down. Because he thought it was a game, he loved doing it as quickly as he could. I still remember his silly grins of pride when he sat down the very second Abby got on the floor.

MOVING FROM LEVEL TO LEVEL

Since every dog is different, there is no way to predict when he'll catch on. Remember Flash, who took twenty minutes to learn to sit at the second level, when we were sitting on the floor? We had to hang in there with him, without making him feel stressed or like a failure, and eventually he got it. After that it was smooth sailing, and he has proven himself to be a brilliant dog. Since he loves learning, and his owners love coaching him, I've taught him many more exercises than the average dog knows.

I've known other dogs that learned to sit in two minutes, but it really doesn't make a difference. Coaching your dog is not a competitive sport; it's a relationship in which you build trust and encourage your dog to become a well-mannered member of society.

Before your dog can learn new behaviors, you just may have to go back to square one, which means returning to sit. If I want my dog to back off, he has to sit first. If I teach him to lie down, he has to sit first, just like he does before I give him his food. Any time he isn't sure what I want, I take him back to his basic foundation, the power of sit. If he does it on his own, that is his way of saying, *I want to understand you, but I need your help. I'm sitting politely, so can you please make it easier for me?*

When you're teaching your dog a familiar behavior, such as sit, in a new place, he may need guidance. If you only train him in one specific room of the house or at the same time of day, he may not be able to figure out that he should repeat that behavior anywhere else or at any other time. Dogs are not good at generalizing, so vary the times and places where you teach your dog to sit, as well as constantly shifting the different levels.

Practice sit many times a day, and include it in your relevance training by making him sit for everything he wants, as if he is saying 'please'. Don't forget to focus on areas – like near the front door, in front of his food bowl, or in the backyard – where your dog is most likely to be excitable and out of control. Comedian and actress Lily Tomlin said, 'The road to success is always under construction.' I don't believe you can teach a dog any skill just by training for fifteen minutes in the morning and in the evening, so practice as much as you can, using any free minute you have. Each time you find ways to make your dog feel successful, you are getting closer to having that unbelievable dog you always imagined by your side.

SUMMARY OF THE POWER OF SIT:

- **The Magnet: use your hand like a magnet and move from his nose to his tail, to get him to sit.**
- **Three levels: teach him to sit when you are:**
  1. Standing
  2. Sitting or crouching on the floor
  3. Lying down on the floor
- **Generalize sit: teach sit in different contexts.**
- **Relevance training: teaching your dog to sit is the equivalent of teaching your child to say 'please'.**

Chapter Ten

Jumping

Whenever I ring a doorbell, and the first thing I hear is a dog barking inside the house, I know I'm at the right place. The door opens, and suddenly I'm engulfed by a tsunami of *dog*! It's a normal, daily occurrence in my life as a dog behaviorist, and I feel so lucky that wherever I go, there is a dog awaiting me. I love the love, but let me tell you, I can do without the jumping.

Jumping is a natural behavior for dogs. They do it when they greet one another and when they get excited. To dogs, jumping is a physical representation of joy, but to human beings, it is bad manners. What if your guest is afraid of dogs, wearing nice clothes, or holding a drink? Jumping can be a big problem, but it is actually quite easy to solve.

For starters, let's understand why a dog jumps on people. There are several different reasons, both social and instinctual. Socially he's excited to see you, and he's trying to get closer to your face. If you knelt down, the dog would not jump above you. He wants to see your face up close, to say hi, and to connect with your facial expression. Instinctually jumping is a necessary behavior for a wolf's survival. When the pack goes hunting, the pups stay in the den with a designated babysitter. When an adult comes home from the hunt, the pups begin jumping and squealing shrill, high-pitched sounds. It looks like a fabulous homecoming, and it is, but it's also much more than that. The pups' jumping and squealing and licking around the adult's mouth causes her to regurgitate the partially digested raw meat she has just wolfed down. The pups eat up what she has regurgitated, and everyone gets fed.

Understanding the instinctual origin, however, does not justify bad behavior in our dogs. We need to retrain them to understand how to behave properly in our homes, which is why socialization is so important. Physically pulling your dog off your guest is not the answer, and neither is asking your friend to correct your dog. That is not a guest's responsibility, it's yours. It's also *utterly* unacceptable to do as some trainers recommend – to lift your knee and plant it hard right into the center of the dog's chest. I've seen a few conventional trainers suggest that when a dog jumps, you should tightly grip his front paws with your hands and step on his back paws. That can cause severe damage to the dog's feet.

I once worked with a handsome white shepherd who had become extremely aggressive. He had started out as a loving, happy dog, but something had gone wrong. It turned out that a trainer had repeatedly stepped on his back paws with heavy cowboy boots whenever he tried to jump. The dog's paws were permanently damaged. By the time the owners realized that this so-called trainer was wounding their dog, his feet were so destroyed that he could only walk for a few minutes at a time. That kind of heartbreaking 'non-training' only teaches the dog to mistrust people – and he *still* won't understand what you want.

Remember that your dog is excited to see you; that's why he's jumping! He's trying to get to your face, where he can see how you feel about him. His joy and desire for connection don't deserve a kick in the chest or the painful crunch of his back paws under the weight of a much larger body. The Loved Dog method of training your dog is far more effective than using physical violence.

NEGATIVE FEEDBACK FOR JUMPING

As we discussed in the treats chapter, a punished dog becomes a stressed-out dog, which can cause unpredictable behavior and make training extremely difficult. Instead, we use the opposite of a reward, which, contrary to popular belief, is not punishment. It is a non-reward, which creates a sense of loss and is a subtle, gentle form of negative feedback – which keeps the dog motivated without pain or fear.

Since your job is to let your dog know if you like his behavior or if you want him to stop, you'll need to use negative feedback to communicate your desires. Praises and rewards only go so far – and just as your dog responds to winning the jackpot, he'll also pay attention if you ignore him.

When I worked with figure skater Scott Hamilton, his wife Tracy, and their two-year-old son, Aidan, I showed them how to use negative feedback to teach their Cavalier King Charles spaniel,

Boogie, not to jump. To start our play-training, I went outside and rang the doorbell. When Scott opened the door, Boogie was so excited to see company, he immediately jumped on me to get to my face. I turned my back, which the dog interpreted as negative feedback, since my face was suddenly unavailable to him. As soon as he stood on the floor on his four legs, not sure what to do, I gave the hand signal for sit. When he sat, I 'made a party' by giving him a jackpot and saying 'sit' with a happy voice and a smile on my face. After a few tries he started to get the picture.

This exercise is so simple and effective, even a child can do it, which I demonstrated with Aidan. 'What do you do when he jumps?' I asked. This sweet blond-headed boy turned his back to show us. Then we 'made a party' for Aidan, in the exact same way we did for Boogie, minus the doggy treats. Each time Boogie jumped, I repeated my behavior. I turned my back until he was standing squarely on the floor once again. I asked him to sit and gave him a treat.

When I turn away, it is a clear message that I'm giving him negative feedback. Turning my back shows him that I'm not giving him what he wants, but rather I'm taking something away. With a little silence the dog will figure out that jumping doesn't equal attention. I then tell him to sit and give him a treat, and he understands that sitting *will* get him what he wants.

When a person tries to physically inhibit a dog from jumping by pulling on him or trying to trip him, they are making a huge mistake. Don't try to stop your dog from jumping. Let him jump at first. How else will you show him that his behavior is unacceptable in your world? Let him jump and let him experience the disappointment of your turning away from him.

REWARDING GOOD BEHAVIOR

Patricia's boxer, Henzo, was so excited when I rang their doorbell, he acted like an exuberant wild man! As soon as the door opened, he jumped on me repeatedly, as if I were his long-lost friend. Henzo was hyper because his only exercise happened when he got Patricia to chase him around the house. I needed to give him an outlet for his energy, so instead of teaching him to sit, lie down, or back off, I engaged him in a game of tug-of-war. Henzo was thrilled to play a new game, and after he had used up some energy, he was much more content. Now Henzo could focus.

At first he jumped on his owner to get a treat. It was as if he was demanding, *Mom, give it to me!* But he would only receive a treat when he sat. I actually encouraged Patricia to re-create the behaviors that had previously caused Henzo to jump, but now all he did was sit to get his reward! Now we were addressing his needs without his having to revert to bad behavior, which is what teaching manners is all about.

I suggested that when Henzo sat in Patricia's presence, every few seconds she could acknowledge his good behavior with a smile and by saying, 'Good sit!' From time to time, she could give him a treat. Once he really understood that he should sit instead of jump, Patricia could switch to random rewards, which would keep Henzo alert and interested.

GUESTS

The best way to teach your dog not to jump on guests is by getting some 'fake' guests to help you re-create a real-life situation. Your fake guests need to know why you invited them over and what they are supposed to do: turn their backs and ignore him when he's jumping and not push the dog away. Any casual social gathering will work. You can even use your training as an excuse to get people together to watch a football game or Thursday-night TV – the rowdier, the better!

During commercial breaks, ask each of your friends to go out, ring the doorbell, and wait for you and your dog to escort him back in. Don't forget to use beverages and snacks as rewards for your generous, dog-loving friends, with a special jackpot for the best performers! Ask the mailman or the neighbors to be actors, and re-create the jumping situation instead of avoiding it. Embrace the challenge head-on.

When I played the fake-guest role for Alicia and her dog, CJ, I purposely rang the doorbell a few too many times, knowing this would cause CJ to get really excited. When he began to jump on me, Alicia politely asked me to turn around. When I did that, CJ realized he was being ignored. His next choice was to turn to his mom, who asked him to sit. When CJ sat and waited, Alicia 'made a party' and fed him some yummy treats. I went back outside to repeat the sequence. In record time CJ started getting it right.

Please don't worry about trying to get your dog to sit and stay *before* you open the door for the guests. Let him jump so he can experience the negative feedback. Soon he won't even try to jump on people because all the pleasure comes from you. Remember that your dog needs lots of acknowledgment from you so that when a source of excitement arrives at your door, you'll be able to hold his attention and keep him from jumping all over your guests.

Beware of a pattern that could etch itself into your dog's mind: jump, sit, get a treat. That's a trap that you must be careful to avoid. If you recognize that kind of pattern starting to form, ignore him until the light goes on in his active little brain. The dog must learn to sit without jumping first. In the beginning you should reward him every time, but once he has learned that sitting is the right alternative to jumping, he won't always need a treat.

SUMMARY OF JUMPING:

- **Negative feedback: turn your back on your dog if he jumps.**
- **Reward the sit: ask him to sit instead of jumping and reward the behavior.**
- **'Fake' guests: ask designated guests to turn away as you capture your dog's attention and restore his manners.**

Chapter Eleven

The Best Wolf Game

Because wolves hunt in packs, they are completely interdependent and can't afford an injury to any pack member. How intelligent they are to have devised a way to challenge one another and establish the pecking order in the hierarchy without causing any harm: games! Let's look at the three games that wolves and dogs play as young as two weeks old. In order to survive in the wild, wolves must be superb hunters, and they hone their skills by:

- **Chasing the prey.**
- **Wrestling it to the ground.**
- **Tugging at the prey to get a fair share of the meal.**

Although games are a fantastic way to teach, I do not recommend for you and your dog to chase or wrestle together. Chasing is a bad idea because it teaches your dog to run away from you, which will handicap your dog's ability to learn to come when you call her. Also, dogs are so much faster and more agile than we are, and they will always win. It will give her the idea that she is alpha, because she can outrun you.

Wrestling is not a good game to play with your dog, either, because when dogs wrestle, they use their teeth. That can be dangerous because a dog may try to initiate wrestling with a stranger or a child, and the results could be disastrous.

TUG-OF-WAR

The third game – tug-of-war – encourages healthy competition among the members of the pack. It represents the last stage of the hunt, when dogs establish social rank by challenging one another for food. The leader of the pack will always end up with the largest and tastiest share. Tug-of-war is the only one of the three wolf games you should play with your dog because it's the only one you can win – and you *must* end up with the prize.

I first witnessed tug-of-war while watching wolves in the wild, and I didn't immediately realize it was a game. I saw a young wolf grab a tree branch and start gnawing on it. In a minute another young wolf tried to take it away, and the game was on. At the time I thought they were fighting, especially when the second wolf tried to tug the branch out of his sibling's mouth and was left with a mouthful of leaves. He growled and walked back toward the winner for a few seconds to see the first wolf turning his body, as if to body block him from reaching the branch. I soon realized they were playing, not fighting, and they reminded me of children. I've seen the same scenario a million times, when two kids want the same toy and one does his best to block the other from taking it away.

I decided to try tug-of-war with my own dogs and they couldn't get enough of it. I was surprised that the dogs were so much happier and better behaved after the game than before it. I started by using a tree branch with my dogs, but a plush toy or a rope works better. A squeaky toy is best because the sound enhances the dog's enjoyment, and it can help to get his attention.

Contrary to popular belief, tug-of-war, if played correctly, will *not* make your dog aggressive. Rather, it will calm him down by giving him a proper outlet for his need to challenge, while re-establishing *you* as the leader. Imagine taking a hyper child to a karate class. The child must understand that there is no kicking or fighting in school, but at five o'clock, he'll get a controlled environment in which he can safely kick and punch to his heart's content. In the classroom, there are guidelines and directions to follow, so no one gets hurt, and the child knows that if he refuses to follow the rules, the teacher won't let him participate. He'll want to behave, so he doesn't miss out on this chance to use up his extra energy and have fun with his friends.

If tug-of-war is played correctly, it can actually establish peace, even though you might feel uncomfortable with the fierceness in a dog's play and the sounds that he may make. Think about football and hockey – both sports look extremely aggressive and combative. The players psyche one another out with angry warnings and glares that imply they want to kill one another. But they end the game with a handshake and a friendly 'Good game!' Then they go off for a beer to celebrate or lick their wounds, depending on whether they won or lost.

Recently I met the owners of a frisky little Jack Russell terrier named Ernie. They loved him dearly, but they were beside themselves with frustration. They desperately wanted to incorporate Ernie into their lives, but he was exploding with so much energy, he couldn't sit still for more than a few seconds. Like a toddler with ADD, he was so hyper that no one could get his attention long enough to teach him anything.

I took a rope toy and held it out for Ernie. Ernie grabbed onto one end with his teeth and stared at me. He couldn't believe I was really initiating this fabulous game with his favorite toy. I gave the rope a short tug, and that was all it took. His amazement shifted to determination, and he bit down harder. As we pulled and tugged together, I saw the wolf inside of him staring back at me with a look of delight that I've seen many times before when I played this game. His eyes opened wide, as if he were telling me, *Thank God. Finally. I really needed this.* When we finished a series of tug-of-war games, Ernie was spent, content to lie down by my side quietly.

Under normal circumstances, tug-of-war will calm your dog, but I must offer a word of caution: you need to make sure you're not playing tug-of-war with a highly aggressive dog, or you could get seriously hurt. This book is meant to help you deal with manners as well as some common behavioral problems. It is not intended to solve severe aggression problems, which should be professionally assessed to determine which kind you are dealing with. In order for the aggression to subside, certain things need to happen simultaneously in the dog's environment while you start the process of behavior modification. The truth is that there is no quick fix for aggression, so be alert and make sure you're not in danger when you start a friendly game of tug-of-war.

THE RULES OF THE GAME

Rule #1: You must always initiate the game.

When Duke was rescued from his abusive past, no one had played with him at all, and he had no idea what a toy was. I offered Duke the end of a squeaky plush toy, and he stared at me like I was a nut. I started by moving the toy along the floor, as if it were a live animal, squeaking it the whole time. I paid attention only to the toy and pretended to have fun with it. Then the moment Duke put his mouth on it, I let go and let him win as if he was incredibly strong. Everybody likes to win, so when I reinforced that behavior, he tugged harder on the toy and held on to it much longer.

Recently I worked with a white boxer puppy named Paco who, like Duke, did not know that tugging on a toy could be so much fun. I had to wake up Paco's wolf instincts by putting a toy on the ground and squeaking it to get his attention. I did *not* shove it in his mouth. That's hardly an invitation. Would you consider it an invitation to lunch if someone shoved a turkey sandwich in your mouth?

Because dogs have a natural instinct to look for small prey on the ground, I placed the toy on the ground beside Paco. As soon as he put the toy in his mouth, I grabbed the other end. Within minutes my playmate made tougher and tougher grabs for the toy. After I let him win a few times, he was hooked for ever – exactly the kind of addiction you want for your dog!

Once Paco got it, I began playing with him by moving my arm in a rhythm, using long, slow movements from side to side, with both of us holding on to the toy. He scooted along, holding the toy in his mouth with all his strength. Then I gave him a short burst of tugs when he was directly in front of me, finally letting him win. He held the toy victoriously in his mouth, but he did not run away from me. I never took the toy away from him, and so he trusted me. At this stage, I let him win ninety percent of the time, to build his confidence and his desire to keep playing.

Rule #2: You must win ninety percent of the time. If you win all the time, you'll lose your playmate, because nobody wants to play a game they always lose.

Once a dog is hooked on the game, I only let him win ten percent of the time. This establishes my alpha position, but it still gives him hope that he can win, too. Remember that your dog must always see you as leader of the pack. Otherwise you are inviting chaos into your home and into your relationship with your dog. Some people use brute force and cruelty to demonstrate dominance, but that will only make your dog insecure and resistant. It's much better to establish your position as boss by using the language that dogs have used for thousands of years. Playing the right wolf games not only sends a clear message that you're the leader, it's a great bonding opportunity for you and your dog, and it's fun, too!

Rule #3: Drop it!

Your dog needs to drop the toy when you ask him to let go of it, but this behavior does not come easily to a dog, since it goes against all of his instincts. I have a couple of suggestions. First, you can use treats or food to entice the dog to drop the toy in favor of a more delicious-smelling item.

Moby, the three-year-old Ridgeback mix, was a wild man with no rules and no manners whatsoever. He had taken over the leadership role in his world, and it got to the point where he became aggressive when anyone tried to touch his food or his toys. I didn't want to agitate Moby, so I used food to coax him to drop the toy. He put up a fierce-looking fight to hang on to the toy, and I felt like I had been shaken and tossed around by a crocodile. Moby was so adamant about keeping that toy, when I tried to coax him with a yummy liver treat he usually adored, he couldn't have cared less.

In the middle of a high-stakes tug-of-war, not even my go-to liver treats were exciting enough! I needed an even better gold treat to urge Moby to open his mouth, and a piece of hot dog finally did the trick. I needed to teach Moby that dropping his toy was pleasurable because he would get it right back, along with a treat. Up until now he had associated the word 'drop' with a feeling of pain. Now, it tops his list of pleasurable things that he and his owner can do together.

The second option involves touching the dog, a rare exception to the no-touching rule. I say, 'Moby, drop' twice, and if he still holds the toy, I use my middle finger and thumb to create what I call 'a bridge' and I place it over his muzzle.

I gently push the skin on the sides of his upper lip down into his mouth, into that space right behind the canine teeth. When he tries to bite down in an effort to get a better grip on the toy, he will feel his skin touching his teeth and he'll open his mouth, at which point I secure the toy. I use the gentlest pressure, causing no pain, and this 'drop' behavior demonstrates to him that I can get the toy away from him without using force. That puts me in the leadership position by holding the resources.

Rule #4: Sit!

The moment the dog drops the toy, I ask him to sit, and then I give him the toy back so the game can continue. I say, 'Take it!' when I give it back to him, so the dog will learn not to be afraid of dropping his toy. He sees that he'll be getting it *right* back. The dog learns that he won't forfeit pleasure by letting go of his toys, so he'll be happy to do it without aggression.

The sequence is: drop, sit, take it! I perform the sequence at lightning speed at first, because I want the dog to learn the pattern. Until he understands the pattern, please do not test your dog's patience by making him wait for the toy. You want him to see that his behavior gets an immediate reward.

Rule #5: You end up with the toy.

At the end of the game, you must take the toy and put it away, out of the dog's reach. You should stop way before your dog loses interest, and always leave him wanting more.

Many people think they know how to play tug-of-war with their dogs, but, in fact, they don't. They roughly snatch the toy out of the dog's mouth, turning it into a game of physical strength instead of a game of control. Then, when the game is over, they walk away, leaving the dog with the toy. This distorted version of tug-of-war will give your dog the message that he is more powerful than you, because he is left holding the resources. No alpha wolf or dog in his or her right mind would ever leave valuable resources to a lower-ranking animal!

NO MORE

More than one client has asked me in deep frustration, 'How do I get the batteries out of my dog?' Well, there is no off switch, but there are two little words, 'no more', that can convey a great deal of meaning to your dog.

After I taught Moby tug-of-war, I put the plush toy that we were using on top of a large dresser out of his reach. Then I told him 'no more' with my arms empty and my palms facing up. If Moby were a wolf, this would have been our battle over the spoils of a successful hunt. Tugging and pulling on each side of our kill would have determined which of us ended up with the food. But instead of fighting over the toy, I say 'drop' and then 'no more'. That puts me in the position of power and strength. If I leave the food – or, in this case, the toy – on the floor where the dog can get it, I lose. If I put the toy away and say 'no more', my dog and I both win, because I reinforce my position as the leader, and he learns good manners.

'No more' means just that. It's not that I am out of treats or toys. I have both, and I also have plenty of time. But when I, as the leader, decide that the game is over – whether it's a ride, a walk, a swim, or even a long game of hide-and-seek – I release the dog by telling him 'no more'. My boys know the meaning of those two words: I am either too busy or have played with them long enough. I don't appreciate being nudged and pushed around, so when I say 'no more', my dogs come and lie down near me. They may get bummed, but no one can have everything he wants all the time. People often use the word 'okay' as a release word, but that expression is too generic for a dog because we often use it in casual conversation. 'No more' will reinforce that you are the leader, and you say when you're finished – on your time, not his.

SUMMARY OF TUG-OF-WAR:

- **Rule #1: You must always initiate the game.**
- **Rule #2: You must win ninety percent of the time.**
- **Rule #3: Drop it! The dog must learn to let go of the toy.**
- **Rule #4: Sit! The dog must sit before he gets the toy back.**
- **Rule #5: You must end up with the toy.**
- **Becoming the leader: end tug-of-war with 'no more' to establish yourself as the leader.**

Chapter Twelve

The Power of Toys

While Patricia looked on, I dug into the large basket over-flowing with doggy toys in the center of her living room. I grabbed a yellow stuffed lion from the top of the pile and called out 'play toy'. Henzo the boxer watched us, wondering what we were up to. He hadn't so much as sniffed that basket of toys in months, but each time I took one from the pile, he became more interested, not so much in the toys themselves, but in what I was doing with them.

'Play toy,' I said once more, as I took a teddy bear out of the basket. That was followed by two more teddy bears in different colors and sizes, and a few more colorful plush toys – *all* play toys. Henzo's well-intentioned mom had bought him a load of play toys, but she never taught him how to play with them! Patricia began to laugh as I continued to go through Henzo's abandoned toys. Each one I grabbed was a play toy, except for one measly little chew toy at the bottom of the pile called a Nylabone.

Dog toys fall into two categories:

- Play toys: to play with you
- Chew toys: to play with by himself.

PLAY TOYS

Have you ever come home to find a stuffed animal's body parts strewn all over your floor? There are no toys in nature, so unless taught otherwise, a dog does to a toy what any wolf would do – he 'kills' it. In order to stop the destruction and keep his toys 'alive' a little bit longer, you have to teach your dog appropriate games such as fetch, catch, and tug-of-war that will take care of his innate wolf aspects and needs.

My favorite play toys are soft plush toys that I use for tug-of-war. My dogs and I also love using soft Frisbees and tennis balls. Any time you buy a toy for your dog, make sure there are no buttons or other objects that he can swallow, or any other parts that may injure or choke your dog.

CHEW TOYS

Chewing is often a big problem for dog owners. When left to their own devices, dogs can become fashion designers, tearing out seams and rearranging hem lengths with their mouths. Some dogs become interior designers, reworking the curtains with their teeth until they look like hula skirts. My dear friend Wrenn came home one evening to find that her entire home had been redesigned by her little Sheltie – not fun, and very expensive to fix!

Chewing is a natural behavior for a dog. That means you cannot stop him from the activity, but you can redirect it to a proper outlet – chew toys. A dog may chew on a play toy, but that doesn't make it a chew toy. Chew toys are things your dog can chew on (in some cases consume) without hurting himself. Some popular chew toys are 'bully' sticks, pig ears, Greenies (but make sure she doesn't choke on the small stub that is left when the bone is almost gone), cows' hooves, pieces of antler, and real animal bones.

If you prefer real animal bones, go to the butcher and get a heavy duty knee or hip bone. Cook the bone in the oven for about forty-five minutes at 400 degrees Fahrenheit or 200 degrees Celsius (to kill E-coli and other forms of bacteria) and allow it to cool completely before you let your dog chew on it. Never, ever give your dog chicken bones. They are brittle, and they can cut the dog's throat on the way down, causing choking or internal bleeding.

In some of my clients' homes, I have to be careful not to trip on a toy when I walk through the front door. It seems like they've gotten their dog every toy ever made, but he lost interest in all of them. You have to let each toy work for you! I suggest you dole out chew toys to your dog, one or two at a time, to keep them exciting. Otherwise, he'll take them for granted and find your most expensive high heels and turn them into Jimmy Chews!

STUFFABLE CHEW TOYS

I like to use chew toys that don't need to be constantly replaced. I particularly like hard, cylindrical rubber toys with a big opening at the bottom for stuffing treats inside. A smaller hole at the top prevents your dog's tongue from going in there, creating a vacuum and getting stuck. For immediate gratification, stuff the toy with small pieces of food. When you first teach a dog how to play with a treat-stuffed chew toy, let a small piece of food stick out, so he realizes that something is in there.

To keep your dog stimulated and occupied, fit in a big piece of a great treat that will not come out. You can do this by squeezing the toy into an elliptical shape and stuffing it. It will return to its original shape when you let go. This will drive your dog nuts while he tries to dislodge the treat, giving him less time to redecorate. Try stuffing your dog's breakfast or dinner in a chew toy, getting him to work for his kibble, just like we have to do. Use something healthy that your dog likes, such as meats, yummy dog treats, fruits, and veggies – anything we discussed in the treats chapter.

Most dogs get bummed out when you leave the house, but if you stuff a few chew toys with gold treats when you close the door behind you, she won't mind a bit that you're gone. All she'll want is to go find those toys, and now, you suddenly become less important in her eyes. In this way, you won't leave a depressed dog in your house, upset that you're gone and looking for some nice furniture to chew into modern doggy art.

To make the game even more fun and challenging, I like to hide several chew toys around the house, sending the dog on a treasure hunt. While you're hiding the toys, ask your dog to stay, and have a helper hold your dog or use a leash temporarily to keep him back. Put the toys in different places around the house, leaving a couple in plain sight. Keep the dog in the stay position, praise him for staying, and then go to the door. As you're closing the door to leave, release him by saying, 'Take it!'

Instead of whining and crying that you left, he'll be thrilled you're gone, so he can start searching for those toys with the surprise treats inside – and he'll associate your departure with pleasure, not pain. Sometimes, I sneak a peek inside and watch my dogs in action. They are so focused on treasure hunting, it really makes me smile.

CATCH

Catching a ball is a favorite game for many dogs. Since you're the one who keeps and throws the ball, playing this game will strengthen his association with you – not just as a leader, but as a *fun* leader. When I wanted to teach Henzo how to catch a tennis ball, I started by gently aiming it at his mouth to make it easier the first time around and he almost got it. To help out his eye–mouth coordination, I took a marker and darkened one side of the ball. This time, after the ball bounced once, Henzo caught it in the air and we 'made a party' for him. We wanted him to catch it before it bounced, but since this was an accomplishment in the right direction, he got rewarded. The next time I threw it toward his mouth, Henzo caught it! 'Catch, catch, good catch!' Patricia and I sang out simultaneously as we high-fived each other. Henzo was one proud dog, and Patricia was one proud owner.

Once your dog is really good at catching the ball, you can teach him to catch a Frisbee. Gently aim the Frisbee at his mouth like you did with the tennis ball. When you're teaching catch with a Frisbee, start by throwing it short distances. Later you can toss it through the air and teach the dog that coming back to you with the disk is a great thing. After all, you are his best cheerleader. Share in the joy of his accomplishment and don't take the Frisbee from his mouth. Let him drop it. Clyde is a champion at catch, and he can hardly wait to bring the toy back to me for another round.

FETCH

I tried throwing a tennis ball with CJ, tossing it only a short distance so he could succeed right away. When he returned to me with the ball in his mouth, I petted him on his back, keeping my hands away from his mouth so he wouldn't think I was going to take the ball away from him. 'Good fetch!' I told CJ. When he dropped it beside me, I picked it up, gave it a couple of good squeaks to get his attention, and I tossed it for him again.

People make a common mistake with fetch, when they take the ball away the moment the dog brings it to them. From the dog's point of view, he is like a wolf back from a hunt, and he is bringing the leader a gift. If I take it immediately out of his mouth and throw it away, he'll think, *What a stupid leader!* So I pet him, saying 'fetch, fetch, fetch' and he realizes that I am not a threat, that I am praising his gift, and I will not insult him by instantly throwing it away.

Recently, I was called to the set of the hit TV comedy *Scrubs*. The actors and crew get to bring their dogs to the set, and they all hang out with one another. But there were certain behavioral situations that needed some rearranging. Stella, a very sweet pit bull belonging to the producer, Randall Winston, refused to return a toy when they played fetch.

'When Stella brings the toy back to you,' I said, 'what do you do?'

'I take it from her and put it away,' Randall said.

I waited a moment until the light dawned in Randall's eyes. Of course Stella didn't want to give her toy to her owner if that meant the game was over. I observed Stella for a little while and noticed that of all the daybeds that were strewn around for the dogs, Stella had a favorite. I sat on her favorite bed, took the fetch toy, and I threw it for her. After she got the toy, she automatically came running over to the bed she liked, and I said, 'Fetch, fetch, fetch' in my best doggy singing voice, and I scratched her playfully by her tail. I did not take the toy away. When she dropped it to lie on her bed, I threw it for her again. She brought it right back and dropped it on her favorite bed, where I just happened to be sitting.

After she had returned it a few times, I took the toy away from her, quickly put some honey and peanut butter on it, and let her lick it off. Now the toy was even better than before! Stella became a champion fetcher in just a few minutes, excited to return the toy so I could apply more of the tasty treats. Shortly after, I turned the treats into random rewards, and Stella continued to return the toy to me. In less than half an hour, Stella had become a champion fetcher, happy to run after the toy, catch it, and return it to me. In the end, I said, 'No more' and kept the toy. Stella was not afraid she would never see her toy again, because I had set a new pattern in her fertile little brain. Bringing back the toy was associated with pleasure, not pain, and now Stella could not get enough of such a wonderful game.

If your dog isn't interested in fetching toys, don't despair. Make a list of the times when your dog is at his highest level of joy. It might be when you take out the leash for a walk, during feeding times, or when you and other members of the family first come home. Start including a squeaky toy at those times, throwing it in the air and bouncing it, showing your dog what a good thing that toy is. Ignore the dog unless he shows interest in the toy, then build on that little by little. It may take a little while. Once your dog becomes obsessed with a squeaky toy, it will become a remote control for him. As soon as he hears the squeak, he'll come running to your side.

SUMMARY OF THE POWER OF TOYS:

- **Play toys and chew toys: give the dog only one or two toys at a time. Stuff chew toys with gold treats, so your dog will keep busy treasure hunting.**
- **Build trust: don't take the toy away from him.**
- **Obsession: get your dog obsessed with catch and fetch games to capture his attention.**

Chapter Thirteen

Come

The best way to call a dog is to bend down in a posture that imitates a doggy play bow, and call out his name. This kind of bow is wolf sign language that implies an invitation to play (it's also the logo of The Loved Dog). It's a good idea to use the dog's name instead of just the word 'come'. What if you're in the dog park, and several owners call their dogs to come at the same time? Your dog needs to hear his name, so he will know that you are directly calling him. Nobody gets confused when I call out, 'Come, Clyde' or 'Come, Duke'.

Sometimes it takes a while before your dog gets enthusiastic about coming to you, and that's okay. Building a relationship with your dog takes time, so be patient. You want to see him wagging his tail so hard that when he shows up, his whole rear end is swaying out of control! I always say that every dog has a tail to tell.

Erin's dog, Honey, was completely disinterested in coming when she was called. Erin repeatedly asked her to come, but Honey would just lie on the grass, half asleep, basking in the sun. Not even a gold treat would get her to move. Since Honey was ignoring Erin completely, I suggested that she stop feeding Honey from her bowl for about a week. Erin fed her out of her hand only, and only while practicing come. This made it not a luxury but a necessity for Honey to come when she was called. Erin practiced for a week, and when I returned, the difference was remarkable. 'Honey, Honey, Honey,' Erin called. An orange streak came flying toward us so fast, I couldn't believe it.

When Alicia was teaching CJ to come, she used her voice really well, but she held out her hand with a treat, broadcasting her intentions. It was like a bribe instead of a reward. The idea was to make her voice enough to lure him. When he arrived to greet her and found a surprise treat waiting, what could be better? If you've made a solid connection with your dog, he'll respond to your invitation and will show up with his tail wagging his entire body.

HIDE-AND-SEEK

Hide-and-seek is a useful exercise for getting your dog to come when you call him. This game speaks to both the wolf and toddler aspects of the dog. For the times he loses track of your location while running around with his friends in the park, he needs to learn how to find you.

I asked Patricia to stand a good distance away from me, and we would take turns calling her dog. 'Henzo, Henzo, Henzo,' I called out in an excited tone of voice, like I had the best surprise in the world for him. He rushed over, and only when he was right beside me did I introduce the word 'come' over and over, while I gave him a jackpot of gold treats. When I said, 'No more, no more,' I also did the hand signal with both my palms facing up. Then it was Patricia's turn.

She called his name, but she said it only once, and her voice was so weak and quiet, I could barely hear her. I encouraged her to speak up and put some excitement in her voice. When she did that, Henzo rushed over to her and was rewarded. She said, 'No more', and it was my turn again – but I had changed locations. Henzo expected me to be in the same place where he saw me last – surprise! He had a great time finding me each time I moved, not to mention getting all those gold treats.

I was teaching him to focus and as much as he loved the treats, finding us was even *more* fun. Dogs love a challenge, especially if they know the end result will be a pleasurable one. When you make coming to you a pleasure, and you acknowledge him for it, he will never be able to get enough. When he absolutely loves to come to you, then you can raise the bar and ask him to sit before you give him a random jackpot.

THE DOG PARK

'Whenever I call Oliver at the dog park, he won't come,' said Vern, the frustrated owner of a cocker spaniel. His little black dog loved the dog park so much that he never wanted to go home. 'I don't know how to get him to come when I call him without making a scene.'

'Does he come when you call him in the house?' I asked.

'Yes, he's perfect. I only have trouble at the dog park,' said Vern.

'Let's go back to his first visit there,' I said, putting on my detective hat. 'What did you guys do?'

Vern didn't hesitate. 'I brought him to the park, and he played with the other dogs while I watched.'

'Then what?'

'I called him, he came, I put him in the car, and we went home,' said Vern.

'How about the next time?' I asked.

'It was the same,' said Vern. 'But he didn't come so quickly.'

'And you're wondering why Oliver stopped coming to you?' I said. 'Why would he come when he knows that coming to you means the fun is over? Do you see the pattern you programmed into your dog's head? You call him, he comes, and you take him away from his pleasure.'

'Oh,' said Vern. 'I didn't realize what I was doing.'

The sad truth is that many dogs associate coming to their owners with pain, not pleasure. For most dogs, the doggy park is the greatest place in the world. Where else can they play with their own species, chasing, wrestling, letting their untamed wolf aspects out? Then, when they are in absolute bliss, they hear their names being called. *The fun is over,* they realize. *If I go back to my owner, he's going to take me away from my greatest pleasure. I'll have to go home and be bored. I think I'll pass.*

If you take your dog to the dog park, I suggest you call him off and on throughout his playtime to give him treats and water. Then tell him, 'Go play!' You'll be teaching him the pattern that coming and checking in with you is a good thing. Use random rewards, and he'll always wonder if returning to you will mean a jackpot of gold treats, a possibility he can't ignore.

SUMMARY OF COME:

- Use a play bow: teach your dog to associate coming with pleasure.
- Hide-and-seek: turn come into a fun, challenging game.
- The dog park: use random rewards to establish a pattern.

Chapter Fourteen

Backing Off

'Stedman, come over here. You have to see Gracie!' Oprah called out excitedly.

We were on the grounds of Oprah's sprawling estate in Montecito, watching her three-month-old dog backing off like a pro. The smallest of the three dogs and dubbed the runt of the litter, Gracie was so small and fragile that when she was born, she needed to be hand-fed every two hours, which Oprah did willingly throughout several days and nights. It wasn't clear if she would make it, but Oprah refused to give up. Today, Gracie is small in stature, but she's a brilliant, gregarious, and loving girl. Oprah's dedication literally saved her life. Now Gracie was showing off for her 'mom', demonstrating the latest and most exciting new exercise that she had learned.

As Stedman approached the porch area, Gracie could not have looked more proud. Stedman gave Oprah a big hug, and we all watched Gracie together. 'That's my girl,' cooed Oprah. After a very short play-training session with me that morning, Gracie had taken to backing off like Oprah's other dog, Layla, had taken to the pond. Stedman gave Gracie a pat on her head and praised her in a happy voice, amazed that the little munchkin was not only alive, but was actually performing the tricky off exercise.

Like Oprah and Stedman, many people come unhinged when they see a tiny thing, like Gracie, backing off when we ask her to. Everybody has seen a dog sit, lie down, and heel, but they are stunned to see a dog back off. This is because in conventional training, off is not considered a basic command. In The Loved Dog method, however, sit, come, and off are three exercises that I find to be useful on a daily basis. What if the dog is about to step on something dangerous during a hike or on broken glass in the kitchen? What if he decides to demand food from your dinner guest? What if a guest doesn't like being sniffed or is afraid of dogs?

When I teach off, I start with sit. As with almost all of the play-training exercises in The Loved Dog method, sit is the foundation, or Square One. Scout, a Westie, was good at sitting, so I didn't give him a treat for that. Instead I held a treat at his nose level, almost touching him, and repeated the word 'off' as Scout tried to steal the treat from my fingers. He nibbled at it, but I held my hand as if it were set in stone. It was important for him to realize that my hand would remain stationary, and he was the one who had to move. My hand didn't even wiggle or shake until Scout made a motion away from me, however subtle, with any part of his body. When Scout moved his head backward an inch, I immediately 'made a party', singing 'off' in a celebratory voice and giving him a jackpot.

Beware of two common mistakes when teaching off to a dog. First, don't hold your hand too far in front of the dog's nose. This will cause the dog to come forward in order to get the treat, which teaches him to do the opposite of backing off from you.

The second mistake is jerking your hand back when the dog starts sniffing or nibbling on your closed hand. This will defeat your purpose since the dog will follow your hand. Once you put out your hand, you've made your commitment to the off exercise. I want you to teach your dog that lunging and grabbing things from people does not get rewarded, especially if he uses his teeth.

Scout quickly learned that backing away from the thing he wanted would – surprisingly – result in pleasure. The next time I held a treat by Scout's nose and said 'off' he moved his front legs back and got a jackpot! At first it may seem like the dog is just nibbling on your fingers without really getting it, but don't be deceived. The dog is thinking about it, and you need to let him work it out. Just keep your hand completely still. Your dog will surprise you by figuring it out and moving his head or a paw back – or maybe even his whole body, like Scout did. Acknowledge any success, no matter how tiny it may be. A great opportunity could be when you first put your hand by his nose and he comes close to you to sniff it. A few seconds later he will pull his head back from its outstretched position, and that's when you want to jump in. Say 'off' and 'make a party' – even if he backed off only half an inch.

Keep in mind that backing off is counterintuitive for a dog, but teaching him to do it anyway is part of the domestication process. Your dog may arrive at your home physically domesticated, but you have to domesticate him behaviorally, and that includes teaching him not to snatch things out of your hands or anyone else's.

When I teach off, I look for a certain progression from the dog. At first he'll probably just move his nose or neck back, which I will reward. After he does it a few times, I raise the bar, which means it will take more on his part for me to give him a reward. I stay silent with my hand in the off position, and most dogs touch me with their noses, as if to say, *Hey, lady, you may not have noticed that I just moved my neck. So for your benefit, I'll do it again.*

I still wait, and he wonders why. I'm actually waiting for him to move a leg backward. Then I'll wait for him to move his tush backward or, in some cases, sideways. When he does, I make a big fuss that was worth waiting for. A pattern will come together in the dog's mind. As soon as he sees that any movement away from you with any part of his body will generate a treat, he'll wonder, *What will happen if I move way back?* A bigger party will happen, or a jackpot!

While he's figuring out what you want, it is not uncommon for the dog to give you what I call 'a buffet' of behaviors that he already knows, for you to choose from. It's so cute, and you will *want* to reward him for trying so hard. But you have to be patient and do nothing. Ignore the behaviors you didn't ask for and reward only the one that you want.

Moby from *Who Gets the Dog?* made great strides with off, even though he moved his body sideways instead of back. He's a goofball of a dog, but since the purpose of off is to be able to instantly remove your dog out of the way of someone or something, I didn't mind too much. If you want to get fancy, you can do what psychologists call 'shaping' once your dog backs off really well. Once he consistently and happily moves away from you when asked, you can start to reward only the straight backward attempts – but that is really for those who have reached an advanced stage of training.

Learning to back off was a huge accomplishment that affected Moby's food possessiveness issues in a positive way. When I coached him, I gave him two choices: he could guard the boring food that was in his bowl, or he could practice off by moving away from the fascinating treats in my hand. He began to realize the good consequences of leaving his food bowl unattended. Not only would he end up getting treats, he would also get a party, as we clapped, smiled, and made him feel like the smartest dog in the world.

THREE LEVELS OF OFF

Just like sit, the exercise off is divided into three levels of difficulty. The purpose of these levels is to get to the point where you don't have to have your hand stretched out with a treat for your dog to back off when you ask him to. Each increase in difficulty level requires increased brain power from the dog. I started Moby with Level One, asking him to back off while I had a treat in my hand and was nearly touching his nose.

When Moby became accomplished at Level One and I was no longer giving him a treat each time, I raised the bar to Level Two. I held the treat in my hand by my belly button, instead of being stretched out in front of me. Moby's owner Abby was pregnant, so I wanted her to be able to hold the baby, whisper 'off' to Moby, and get immediate results.

As soon as Moby understood that, I moved to Level Three, which meant holding the treats behind my back. It didn't bother Moby at all that my hands were out of sight. He had the pattern etched in his smart little head, and he was right there, giving me what I asked for at Level Three. Remember that Level Three is more difficult, because when I remove my hands from the equation, he has to rely solely on his knowledge of the English word, the name you gave to this particular exercise. It may seem like a subtle shift in understanding to you, but there is an ocean of growth between each level, and you must reward the dog for grasping the difference. As you introduce the different levels, please be patient. As with everything in life, the more you practice, the better.

Keep in mind that rewards don't always have to be treats. Make the off behavior relevant by asking him to do it before you serve him his meals, or ask him to back off before throwing him the tennis ball. After each toss, increase the distance he must go before you throw the ball. Now, that's what I call fun!

SUMMARY OF BACKING OFF:

- **Reward any attempt: the subtlest motion backward is worthy of a treat.**
- **Level one: hold the treat in front of your dog's nose. Do *not* move your hand. Let the dog move first.**
- **Level two: hold the treat by your belly button.**
- **Level three: Hold the treats behind your back.**
- **Relevance training: use toys and meals to incorporate off into your dog's daily life.**

Chapter Fifteen

Chilling Down

Dogs love it when we ask them to chill down with The Loved Dog method. It's like a mini-vacation for them. For us, it's about good manners, and it's also preparation to teach them to stay, which we'll discuss in the following chapter. Keep in mind that down and stay are two separate exercises with different meanings. Down means you want your dog to lie down. Stay means he needs to stay right where he is without moving, until you release him. Since the messages 'down', 'lie down', and 'sit down' can be confusing to a dog, use the simplest, most straightforward word, with nothing extra tacked on.

In the old compulsion dog-training methods, 'down' is the command that dogs dread the most. They know that their heads will be yanked down to the ground with a choke chain, and they'll be forced to stay there. Their owners' intense and threatening demeanors will make the dogs feel vulnerable, not only because they are being pushed into a submissive body posture, but also because they do not understand why their trainers are being so unkind. When they are finally released, the last thing they want is to be ordered to lie down once again. A vicious circle can begin if a dog's resistance makes the trainer feel he must become even more forceful.

In The Loved Dog method, teaching the dog to lie down is fun and easy. The down exercise requires absolutely nothing of your dog – in fact, doing nothing is how he succeeds at this technique. Your message to him is, 'Enjoy yourself, chill down, and let me take care of you.' How great is that? Down is so easy to teach, I've seen a two-year-old pull it off. Toddler Aidan Hamilton was a crackerjack at learning to teach his family dog, Boogie, to lie down. I gave him a few treats to hold in his hand, and he had it in no time. We all high-fived him after his success, and both Aidan and Boogie were pretty pleased with themselves.

TEACHING DOWN

Kelly, a stunning Rhodesian Ridgeback, loved to explore. She was sturdy, strong, and exceptionally smart. Her owner, Sam, a contractor, loved taking her to work, where she had a ball running around his building sites, sniffing and exploring. But there were times when Sam needed her to lie down beside him, because large machines moved around and dangerous holes in the ground were exposed. As keen as Kelly was to explore and rush around free, she was not a bit interested in sticking close to her 'dad'. The fun was elsewhere, and the sounds and smells of workers' hot lunches lured her to be anywhere but at her owner's side – until we taught her 'down'.

I started coaching Kelly indoors, where we could set a pattern without the competitive excitement of the outdoors. I asked her to sit, then I crouched down and put a treat directly in front of her nose, preparing to do the magnet method. I looked at the spot on the floor where I wanted her to look, and I made a straight line with my hand from her nose to the floor. As she lay down to eat the treat, I said, 'down', but only when she was already there, not before! I kept placing treats on the floor between her front paws, because I wanted her to stay close to the floor.

A secret to remember: where the eye goes, the body follows.

I look at the floor when I ask a dog to lie down, because the dog looks to see where I am looking. As Kelly complied, I repeated the word 'down', letting her know I was impressed with her. I began the exercise in a crouch, but within seconds, I straightened up and stood at my full height. Each time Kelly looked at me without getting up herself, I put a treat on the floor between her front legs. She was learning that although I was moving around, she had more fun lying down, setting 'down' as a foundation for 'stay', which we'll learn next. I started shifting my weight from leg to leg, and each time she acknowledged me, I gave her a treat. I moved to the next level, taking a few steps to her right side and then a few steps to her left. I moved around in front of her, taking baby steps and giving her treats. I showed her that lying down would be much more rewarding than following me.

As I saw Kelly getting more and more comfortable, I took a little bit longer between giving her treats, all the while smiling and singing, 'Down. Good down.' She did not exhibit the slightest interest in getting up, and she kept her eyes glued to me the whole time – unless she was eating up a yummy treat! Once Kelly had caught on to the pleasure of lying down, it was almost as if she were saying, *I'm not going anywhere, so you can just keep bringing on those treats!* She might as well have been sunbathing in the Caribbean, wearing a wide-brimmed hat and sipping a pina colada.

THREE LEVELS OF THE DOWN HAND SIGNAL

If you remember, the hand signal for sit is done with your palm up, with a motion toward the ceiling. The hand signal for down is done exactly the opposite, with the palm down and moving the hand toward the floor. It's important to remember that when you're asking the dog to lie down, to always look at the floor, not at your dog's face. As always, the three levels increase in difficulty in order to cover all potential situations when you may need you dog to chill down.

- **Level one: touch the floor with your hand.**
- **Level two: keep your hand midcalf until the dog lies down.**
- **Level three: remain standing straight while your hand motion points down.**

NEGATIVE FEEDBACK

If your dog gets up from the down position, like Kelly did a few times, there is no need to correct her. In down, I want you to emphasize the positive, which is the act of lying down. Remember that if your dog gets up, she did nothing wrong, because – deliberately and wisely – you did not ask her to stay. But I still want her to realize that she's missing out by getting up, and eventually she'll *choose* the down position. So when the dog gets up (as they all do during the learning process), I use negative feedback.

Remember how we turned our backs when the dog was jumping? I did the same thing when Kelly got up. I turned my back, acting so disappointed that I couldn't even look at her. I huffed and puffed loudly, clearly showing Kelly that I was unhappy, and she quickly realized that the wonderful flow of love and treats stopped as soon as she got up. *What have I done?* I could see her thinking. *It was so much better when I was lying down.* It is so cute to see how most dogs at this point will come over and sit or lie down for their owners, hoping they can make things go back to the way they were before. Then you can act surprised and 'agree' to continue the fun.

If your dog is very hyper, teach her down after she gets back from a long day of hiking or any other time when she is mellow and relaxed. Since you're homeschooling your dog, you can set up the circumstances and conditions to give her all the help she needs to be successful. Remember that the more successful she feels, the more she'll want to practice. Now you two are working as a team!

FOUR STAGES OF DISTRACTIONS

Teaching Moby to lie down was so easy that he wagged his tail happily throughout the training. When he got up, I turned my back and made exasperated sounds. Moby got it immediately and sat at my feet, showing me he was ready to try again. I was happy to see him self-correcting, so I immediately did the magnet from his nose to the floor. When he lay down, I resumed placing the treats between his front legs.

But how would he do with distractions? I decided to act silly, doing jumping jacks and dancing around, right in front of him. I deliberately rattled the window blinds, making a disturbing noise. Moby watched my antics, but he stayed where he was, content to keep receiving treats on the floor between his front legs. Because I had done enough repetitions of walking away and turning my back, he knew what was expected of him, even when I was acting silly.

As in many of my coaching exercises, I attempt to distract the dog during the down exercise. I use four stages through which he continues to get a variety of bronze, silver, or gold treats, depending on the level of difficulty, with jackpots given randomly.

STAGE ONE: I move around in front of him, a few steps to his right and a few steps to his left, a few steps backward (facing him), and then return to my original position.

STAGE TWO: I do a 360-degree turn, during which I quickly show him my back, still assuring him that I'm not leaving. During the turn, he is my focal point. My goal is to be able to turn my back to the dog and walk away, without him feeling that he needs to get up. At first I do my turns quickly. Later on, I stay a little bit longer with my back to him. After a few seconds, I give him a jackpot, since he stayed put despite all of my turning and moving.

STAGE THREE: I act silly, do jumping jacks, crawl on the floor, toss his favorite toy in front of him, and – most tempting of all – ring the doorbell. The toughest thing for a dog is to sit still when the doorbell rings, so I do it a few times, preventing his favorite activity as a self-designated welcoming committee.

STAGE FOUR: While the dog remains in the down position, I sit calmly on a chair or a couch, or I lie down on a bed. This level is very difficult for your dog, because he'll want to join you, or he'll think your play-training session is over.

When you are done playing, always remember to release your dog with the words 'no more'. You should be the one who determines when the exercise is over.

SUMMARY OF DOWN:

- **Start from sit: sit is the position from which most behaviors begin.**
- **Look at the floor: where the eye goes, the body follows.**
- **Treats: put the dog's treats on the floor in the middle of the space between his front legs, so he can reach them without getting up.**
- **Don't stand still: The three levels of distraction raise the bar.**
- **End the exercise with 'no more'.**

Chapter Sixteen

The Secret of Stay

When I need a dog to stay in one place, I don't want to have to argue with him. He needs to understand the word 'stay'. His chances for quickly learning to stay depend on how well he has mastered down. If a dog chills down and then he gets right back up, he is still complying. Stay is the next step, and when I add that to his arsenal of knowledge, I don't want him to move until I use the release words, 'no more'.

In the down exercise, if the dog gets up, I show him my disappointment and wait for him to sit or get back down to try again. But he is not really defying me, because I did not ask him to stay. When I do ask him to stay, however, he needs to stay right where he is, lying down, until I release him. As with most of my techniques, I teach the dog to stay in steps, ending with a fun game of treasure hunt, which all dogs love.

Stay was a terrific tool for Matt and Abby to help Moby with food aggression. When he wanted food that he could not get instantaneously, he became pushy and demanding. I had to find a way to show him that no matter how far away I stood from him or where I put down his food, he did not have to become a bully in order to get what he wanted. I wanted him to know that he had to follow the rules, and most important, that he can trust people.

I began to teach Matt and Abby how to use stay with Moby. I explained to them that when Moby saw food and he couldn't get it, he actually reacted in fear. *Oh my God,* his brain told him, *I may not get it.* And out came the bully. My goal was to change Moby's programmed response around food. I wanted to teach him to trust that there was food and he would get it, but on my terms, not his – and that he would have fun!

THE SEESAW

When we practice down, the treats are placed between a dog's front legs, where he can reach them with his mouth. In stay, it's the same, but there is an escalation of difficulty, because now I make two piles of food – one between his front legs and another one a couple of feet away. I want the dog to realize that when I tell him to stay, he needs to stay no matter *what* goes on in front of him. He must trust that I'll come back to acknowledge and reward him. I want him to *want* to stay, because it could easily result in fun!

Remember, The Loved Dog training is based on your dog's positive or negative associations. When a child sits in the corner facing the wall as a time-out for bad behavior, it's a shameful and negative experience. But that same child will happily count to a hundred in the same corner in a game of hide-and-seek. Similarly, stay can feel like frustration, or it can be a positive and exciting experience for your dog. The seesaw will help him make it pleasurable.

When I taught the seesaw to Moby, I began by asking him to chill down, and I placed a gold treat between his front legs. Instead of facing him like I did before, though, I positioned my body sideways to him while he was down. He gobbled up the treat, and I introduced him to the stay hand signal by placing my open palm in front of his face like a stop sign. I said 'stay' in a firm, clear voice, but never threatening. Then I put a treat on the floor a couple of feet away from Moby, repeating the stay hand signal as well as the word 'stay'. The fact that Moby did not jump up and charge over to get that treat made his owners gasp with amazement. I did it again, placing a yummy treat between Moby's front legs, doing the stop sign with my hand, and adding another treat to the second pile that was about four feet away from Moby's stay position. Back and forth I went, like a seesaw, putting treats in the two piles, and making sure Moby did not get up.

When I was finished, I put the last treat on the floor between his front legs, and I also gave him a jackpot so he *really* would not want to get up. I looked at sweet Moby, waiting patiently for me to release him, and I said, 'No more. Take it, take it!' with great enthusiasm. Moby got up and rushed over to the pile of treats, eating them with gusto while we clapped our hands, repeating 'Take it, take it' until he had finished them all. This was the first time I'd taught Moby how to stay, and he already loved it.

Down is usually among a dog's favorite behaviors, which makes stay a treasured behavior, too, even though it involves more patience on the dog's part. Matt and Abby were in such disbelief at their dog's instant achievement that Matt suggested we check Moby's tags to see if he was really the same dog they had yesterday! After all, not only did he perform down right away, but he was content to remain patiently in his stay position when he was asked. He did not become aggressive when I wouldn't let him take the pile of treats. This was a *major* breakthrough, a fantastic achievement for Moby, and you could just see how good he felt about himself. Moby knew we were playing a game that offered both predictability and consistency.

During our first stay lesson, Moby showed no signs of aggression. In fact, he was doing so well, I asked Matt to try using the seesaw. The results were nothing short of miraculous. At first I noticed that Matt left out the stop hand signal before he put treats a short distance away from Moby. I didn't correct Matt, though, because it's a lot to learn for the owner as well as for the dog. I wanted to praise Matt for what he was doing right, so he would associate coaching his dog with pleasure. If Matt had put the treat too far from Moby's front legs, or used a threatening tone of voice, I would have kindly guided him to a better way of doing things. Forgetting the hand signal was no big deal.

I change my tone of voice slightly when I'm putting the treat between the dog's legs and when I'm placing a treat a distance away. The treat between his front legs comes with a warm voice, and when I place the treat at the other end of the seesaw, I use a firm voice that cannot be interpreted as an invitation. When using the hand signal for stop, many owners make the mistake of switching to an army training voice. Instead, make sure your voice is matter-of-fact, but also warm and encouraging, so the dog will know you're playing a game and that you're happy with him.

Moby was so patient, I finally directed Matt to give him a jackpot and release him. 'No more,' Matt called out. 'Take it, take it.' Abby and I joined in, saying 'take it' and clapping our hands as Moby rushed over to the pile of treats and ate them while we gave him a well-earned party. When Moby lay right back down, hoping for more, Matt and Abby burst into laughter. He was so cute and compliant, and his owners were completely enthralled with him!

During the initial teaching phase, I give the dog a jackpot before I release him from the stay position. I also give him a jackpot *during* the stay to keep him interested and to show him how well I think he's doing. I never release him when I'm standing a distance away from him, unless I'm leaving the house and letting him go on a stuffed chew-toy hunt. He needs to learn to wait for me to come back to him, no matter how far away I am.

RAISING THE BAR

It was time to take the seesaw a few feet farther away, so I began by giving Moby the hand signal for down. He was so happy that we were about to play again, he got down immediately. I fed him some treats and began the seesaw exercise again, this time lengthening the distance between the dog pile and the out-of-reach pile. I was careful to do the stop hand signal for stay each time I walked away from him. The bar had just been raised, and I needed to make sure my message remained precise, consistent, and clear.

NEGATIVE FEEDBACK

Teaching your dog to stay includes letting him know when he makes a mistake. There are two parts: one is giving negative feedback by making what I call a 'dolphin sound'. The second is taking the dog straight back to where he was before he got up.

When I raised the bar by increasing the distance between Moby and the out-of-reach pile of treats, he got anxious and broke the stay. I immediately let out a strange, ear-piercing, chirping sound, similar to the noise a dolphin makes, to momentarily stop him and show him that I was not pleased with what he was doing. It's an annoying sound, but not angry, like a 'no' or a Level Eight experience. I was using negative feedback with a series of high-pitched sounds that Moby could not help but acknowledge. Once I had his attention, I showed him the replacement behavior that I wanted. I've seen owners yell out at their dog when he is straying from what they want, but they don't follow that up by showing him what they *do* want.

I ask him to lie down and stay, but in order to avoid setting a bad pattern, I don't give him a treat for breaking a position. If I did, he would think that the pattern was stay, break the stay, go back into the stay, and get a treat. It's very easy for a dog to misunderstand what you want him to do.

I used plenty of attitude in my stance and my voice when I made the dolphin sounds with Moby, to let him know that he'd messed up, but I was never nasty. If I'd sounded or looked angry, Moby would have felt stressed and might associate that feeling with stay. Instead, I went to the exact position where he was originally lying down, pointed at the floor (which is like the down hand signal), and Moby went there to lie back down. I immediately softened my urgent posturing and said

'Down' but did not offer a treat or a smile. Then I said, 'Stay.'

I returned to practicing stay as if we hadn't been interrupted, and Moby was a happy dog when he saw the smile back on my face. As his treats got farther away and Moby tried to get up, I made the dolphin sound again, showed him the replacement behavior, and soon enough, he was self-correcting so quickly, I could not have been more proud. Remember, this was a dog who was so possessive of his food, he was ready to attack whoever he thought was keeping him from it. With that in mind, each jackpot I gave him for good behavior was all the sweeter. Even Matt, a military man who had seen action in Iraq, began making dolphin sounds and adding sweetness to his communication with Moby when it was appropriate.

During Moby's first lesson, every time we released him with 'take it' we clapped our hands and 'made a party'. We wanted Moby to feel really good about his accomplishments and to be excited to do more. Over the following weeks, Matt and Abby raised the bar again for Moby by hiding treats in different rooms and releasing him from stay to an exciting treasure hunt. This previously food-possessive dog had changed from an aggressive protector to a carefree pirate on a treasure hunt, thrilled with the game, himself, and his owners!

STAYING WHILE OUTSIDE

Moby was becoming such a good treasure hunter that we decided it was time to move outside and learn to stay, with all the distractions that occur there. We went out to the quiet, tree-lined street, but we immediately ran into a problem. Moby refused to lie down at all, never mind stay.

I wondered what was going on in his mind. He was not acting up, and he did not try to take off. He simply refused to lie down. I knew he wasn't stubborn, defiant, or disobedient, and I soon discovered the simple explanation. Up until that time, he had no experience lying down on anything but indoor carpeting. The asphalt made him feel uncomfortable, so I came up with a Plan B. As usual, solutions are almost always much simpler than we might expect.

I went to my car and took out a fluffy yellow bath mat that I kept in the backseat for Duke and Clyde. When I put it down, Moby sat on it immediately – willing and excited to stay. With Moby lying on his bath mat, Matt performed the seesaw exercise. Since we had already established the pattern indoors, Moby understood exactly what Matt was asking of him. Over time, Moby got used to the asphalt, and his owners made his 'carpet' smaller and smaller, until he started lying down directly on the asphalt without hesitation.

SUMMARY OF THE SECRET OF STAY:

- **Start with down: your dog must love the down position in order to move to stay.**
- **Seesaw: put treats in two separate piles, one between his front legs and the other a few feet away in front of him.**
- **Negative feedback: make startling sounds to let your dog know that he is breaking the position, and show him the correct behavior.**
- **Location, location: generalize the locations for stay, so your dog will have manners wherever you are.**

Chapter Seventeen

Housebreaking

'Tamar,' a man named Brent said to me on the phone, 'I can't housebreak my dog. I thought he was already housebroken, but I've been out more than usual, and he's been peeing all over the house. I've tried everything, and he's so stubborn, he refuses to listen to me. What should I do?'

'What are you doing now?' I asked, almost afraid to hear his answer.

'The usual,' Brent said. 'I stick his nose in it, I hit him with a rolled newspaper, and tell him *no*!'

I cringed. Unfortunately this was 'the usual' because for some reason, many people still think this is the right way to housebreak a dog. It was no surprise that while Brent was unsuccessful at housebreaking his dog, he was certainly successful at *breaking* him, and I wanted him to stop.

'Not only that,' continued Brent. 'The more I shove his nose in it, the more often he pees. He's doing it all over the place just to spite me.'

I had a sneaking suspicion that the problem might be physical, not behavioral. 'Do you take him out regularly?' I asked.

'Of course,' Brent snapped.

'Have you taken him to the vet?' I asked. 'Maybe he has a urinary tract infection.'

'No, no,' said Brent. 'This has been going on for a long time. It's because I'm so busy these days. He's just willful and stubborn.'

Now I was pleading with him. 'Brent, please take him to the vet,' I said. 'It sounds like he has a physical problem.'

'Fine, I'll take him,' said Brent and hung up abruptly. I suspected he wouldn't, so I was saddened – but not completely surprised – that several weeks later, a friend told me that Brent's dog had died. It turned out he had a serious urinary tract infection that, due to neglect, had traveled into his kidneys and his bloodstream. It had killed him slowly and painfully – all while he'd been beaten up for peeing inappropriately.

Contrary to public opinion, sticking a dog's nose into his own excrement and angrily tossing him outside is not the way to housebreak your dog. You don't really have a problem with your dog actually *going* to the bathroom – you have a problem with the inside of your home as the location of the bathroom. Punishing your dog for having an accident only sends the message that the natural act of relieving himself is wrong.

Cruelty can make a dog revert to doing whatever it takes to avoid his owner's wrath, so in order to protect himself, he may start eating the evidence or refusing to go to the bathroom in front of his owner. If he hides his 'business' from you, you won't be able to praise him when you see him going in the right place, and therefore it will be impossible to housebreak him. Nobody wants to find unpleasant surprises in covert places like the back of your closet or inside your favorite shoes.

Ozzy Osbourne told me that he didn't understand why his dogs were peeing all over the house when he kept all the outside doors wide open. The problem was that nobody had ever shown his dogs where they should go. When the Osbournes called me for help, I explained that housebreaking a dog is as straightforward as teaching him any other skill. All you need is to teach the dog two things: to know where the bathroom is located and to 'hold it'.

Where should the bathroom be? Most people immediately think 'the grass in the backyard'. But isn't that where your kids play? Isn't that where *you* like to spend time? The middle of the backyard is not necessarily the best location for your dog's personal toilet. I like to choose a place that is off the beaten path, such as a designated gravel area, under trees or bushes, at the edge of the yard, or on concrete for easy clean up.

USING A CRATE

A crate is the fastest way to teach your dog to hold it. Because dogs are den animals, they love small places, so they will not dislike the crate. The only thing about it they may not like is that they can't come and go as they please, but this is a privilege they will earn once they're housebroken. A couple of my clients didn't want to use a crate, so they used a spare shower in their home instead. They felt this was a kinder choice, but I think a crate is cozier and more suited to your dog. Take a look at where he chooses to lie down when he has the run of the house. It's usually in a small space, like under a table or behind a chair. A crate should be a comforting place, and many dogs like to tuck in there for a nap even *after* they are housebroken.

After his initial reaction to the crate, which may include whining, crying, or barking, the dog will relax and enjoy a little nap. Make this experience pleasant by talking to your dog in a soothing tone of voice, and help him realize that the crate is his own private bedroom – not a jail cell. When your dog is in his crate, give him a yummy chew toy stuffed with a silver or gold treat and feed him at the far end of the crate with the door open. The crate will not represent isolation if it's located in a busy room and your dog sees that there are a lot of good things happening in and around it.

The crate should be big enough for your dog to be able to lie down, stand up, and turn around, but no bigger. Otherwise, he might do his business in one corner and still stay dry and comfortable at the other end of the crate – and not learn to hold it. Always remember that the crate is only there for housebreaking your dog, so please don't use it to punish your dog or for any other purpose.

GETTING ON A SCHEDULE

With some simple rules and a lot of repetition, housebreaking can be easy. First, set up your crate and designate an outdoor bathroom area. Whenever you put your dog in the crate, use a happy voice and give her some good toys. Leave her there for two hours and don't give her water during that time. When you take her out, give her as much water as she wants. Load her up *before* you take her outside to eliminate. If your dog drinks randomly, she will pee randomly, and you'll have a hard time predicting when she's ready to go.

After she waters the backyard, the street, or the bushes, you can let her run around freely in a limited part of the house for a half hour. Because she's 'empty' you can trust her. After thirty minutes, put her back in the crate for the next two hours. If your dog makes a mistake during her 'free time' in the house, make her bathroom breaks a little bit longer. Some dogs relieve themselves in two separate sessions, not all at once.

You should feed her two or three times a day, always at the same time and, whenever possible, *after* the family has eaten. Wait fifteen to thirty minutes after she eats and drinks, and then take her out. Until a dog reaches twelve weeks old, she should eat three meals a day. From then on, feed her twice a day, which will help with housebreaking.

Your dog should get plenty of water, since withholding water can damage the dog's kidneys. It should be available abundantly during meals and every two hours or so, right before you take her out to her designated bathroom area. Let her drink indoors or in the yard with you. Never put water or food inside the crate (a chew toy is okay) during the two hours of 'hold it', because if she fills up her little belly, she won't be able to wait. That's not fair, because you'll be setting your dog up to lose.

The morning is the best time to practice the bathroom routine, since you know that your dog is loaded from holding it all night. When you first get up, take her out to the bathroom area and hang out there. While she sniffs around for that perfect spot, make yourself as invisible as possible. Once she's actually going to the bathroom, repeat the name you've chosen for that behavior ('hurry up' or 'potty'), as if you're singing a lullaby, but don't look directly at her. Don't get excited while she's eliminating, or she may rush over to you to check out what's going on, and she won't complete the process. *Do* get excited as soon as she's done, singing the words 'good potty', or whatever you choose to call it, in an excited tone of voice. Give her a smile and a silver treat. Now you can start your day. At night, make sure your dog gets no food or water for three hours before you go to sleep. You want to keep her as empty as possible, so she can hold it for the maximum amount of time until your morning routine.

My clients ask me what to do if the dog is sleeping when it's time to take her out. Just like with a child, *you* are in charge of the schedule, so you're the one to tell her when to go in and come out of the crate. The most important thing is to focus on performing as many repetitions of this 'in and out' routine as possible each day. When your dog can enjoy a half hour of free time with no mistakes, you can increase it to forty-five minutes and then to an hour. Little by little, according to her behavior, she'll start getting more free time before she goes back into the crate.

How long this process takes is up to you and your dog. The more you stick to a schedule and repeat the 'in and out' routine, the faster it will be. Sometimes it can take only a week, but most of the time it takes up to four weeks for the dog to be trusted to go to the designated bathroom on his own, or to let you know when it's time.

THE HOUSEBREAKING CHART

After a couple of weeks of the crate routine, you may want to create a chart to see if your dog has developed any patterns. That information will help simplify things, because you'll know when to keep her confined and when she can be trusted a little bit more.

Each chart should cover seven days, and everyone in the household should write on it. When did you feed her? When did you give her water? Use the letter *W* for 'water' and *F* for 'food', and fill in the chart according to the day and time. Did you see her go? Was it number 1 or number 2? Was it in the bathroom area or was it a mistake?

Represent her proper bathroom area as a circle and write inside what she did and when. If she made a mistake, write that down in a square. Write down the times of the day when she makes mistakes. By the second week, you'll begin to see your dog's patterns, and you can take the necessary precautions. For example, she may be more prone to making mistakes in the late afternoon – so pay more attention to this time and be sure she gets to go out more than once. When she's on a fairly regular schedule, you can give her more free time since you'll know she's capable of holding it.

RING A BELL

Many owners ask me how they can tell when their dog needs to go, and the truth is that it's different with every dog. To figure it out, you'll have to learn to read your dog's signs and pay close attention to his behavior: does he start sniffing when he needs to go? Does he circle a certain spot? Does he whine or cry?

An alternative approach is to actually teach your dog how you'd like to be notified – even by ringing a bell. I'm not kidding! I worked with my client Erin to show her dog, Honey, how to touch a bell with either her nose or her paw when she needed to go. I started by putting a treat in my hand and holding it right in front of Honey's nose. When she nudged it, I said 'touch' and gave her the treat. I repeated this a few times, until Honey started to get the touch pattern.

Now I wanted Honey to become familiar with the actual bell I'd be using. I used an old cowbell that Erin had in her house, but any bell will do as long as it is somewhat sturdy. I started by dangling the cowbell on a string. When Honey touched it with her paw or her nose, it was party time as I repeated the words 'touch, touch'. After a few more touches and celebrations, it was time to teach Honey to touch the bell while it was hanging on the door.

To help your dog along, you can stuff a yummy treat inside the bell, and when she comes to sniff or paw at it, reward her. When the bell rings, say 'touch' and give her the treat. Since I wanted Honey to understand that touching the bell signified going out, as soon as she touched the bell with her nose or a paw, I immediately took her outside and gave her a treat out there.

For Honey's first attempt, she did not touch the bell, but she got very close to it, which earned her a reward. At first, as with many of my other exercises, I reward every attempt in the right direction. This builds her self-confidence and shows her she is on the right track.

BATHROOM BEFORE A WALK

The training is not over once you have taught your dog to indicate she wants to go out. Now she has to understand that she came outside to use the bathroom. Waiting patiently until she eliminates can be downright frustrating. If you take your dog on a walk *before* she goes to the bathroom, she may take for ever to find the right spot. She will most likely stretch it out even further because she knows that once she goes, the walk will be over. Let her know that she needs to go to the bathroom *first*, in order for her walk to begin.

Scott Hamilton described his dog, Boogie's, maddening behavior when he took him outside: 'I know he has to do his business, but instead he runs out to find a place where rabbits or other animals have been. Then he buries his nose in one place and he won't move for the longest time. And he won't do his business. It drives me crazy!'

Boogie needed to understand the difference between going to the bathroom and going for a walk. I instructed Scott to take Boogie out to the yard first, to his regular bathroom area, and repeat his bathroom word. After Boogie relieved himself, Scott could reward him by taking him for a walk – even if it's just for a couple of minutes.

If your dog refuses to go to the bathroom first thing in the morning when you know he needs to go, you can try to put him back in his crate for another thirty minutes, to make his situation more urgent. And if you really want to show your dog what he's missing out on, take his leash, show it to him, and then take the leash on a walk without him. I know that sounds silly, but it works more often than you think.

MAKING MISTAKES

Albert Einstein said that anyone who has never made a mistake has never tried anything new. Authors, inventors, and artists all agree that making mistakes is part of being successful. I share the same philosophy when training dogs. Don't be afraid to make mistakes, because you and your dog will be trying new things constantly, and he will not always catch on immediately.

No matter how diligent you are, there are mistakes you'll catch right away and some that you won't, and you need to handle them differently. When you catch your dog going potty in the wrong place, make it clear by your tone of voice that you are unhappy. Make the shrieking dolphin sound, and immediately escort him outside. Once in the yard, relax, and wait calmly for him to do his business. When he goes, praise him, give him a jackpot, and let him back inside. This will clearly communicate what you like and what you dislike. I suggest you keep a container of treats by the bathroom area, so you'll always be ready to give a big jackpot for good behavior.

When you don't catch your dog's mistake right away, a different response is needed. Don't bother taking your dog back to the scene of the crime and engaging in various forms of stress-inducing behavior, like yelling, hitting him with your hand or a rolled newspaper, or rubbing his nose in his excrement. When your dog goes in the wrong place, the smell usually lingers, increasing his chances of becoming a repeat offender in that particular spot. To clean up his mistakes, make sure to absorb all of the liquid onto a paper towel, and then use a cleaning product designed to remove pet odors.

SUMMARY OF HOUSEBREAKING:

- **Hold it: use a crate and a regular feeding schedule to encourage your dog to hold it for increased periods of time.**
- **Choose a word: repeat a word or expression, like 'hurry up!'**
- **Bathroom before a walk: make going to the bathroom a requirement before going for a walk.**

Chapter Eighteen

Walking

I was watching the *Dr Phil* show recently, and a woman was describing how she dealt with her daughter's bad grades. She went into great detail about how she nagged and yelled at her little girl to study and focus, grounding her and taking away her phone and TV privileges when she failed. She even refused to talk to her for a few days after a particularly bad report card.

Dr Phil gave this woman a long, steady look and asked in his famous Texas drawl, 'So, how's that workin' for ya?'

I smiled. I don't know how many times I've gone to work with a client who was having trouble walking his dog. Usually, the moment the dog is clipped to the leash, he jerks forward, practically dragging his owner to the ground. Trying to regain some sense of control and dignity, the owner devises elaborate methods of pulling, yanking, choking, commanding, and threatening. The perfect question here always is: 'So, how's that workin' for ya?'

Many owners insist on continuing these useless methods, week after week, month after month, and sometimes year after year, hoping that it will suddenly produce results. Of course, it never does. Choking, pulling, and yanking will ensure only one thing: you and everyone else in your household will dread walking your dog. As a result, he will get inadequate exercise, and he will start turning his bottled-up energy on you, your guests, and your home. This is a recipe for disaster for everyone, particularly for your dog. Exercise is one of a dog's seven basic needs, and walking is a great form of exercise for both you and your dog. It's also easy and fun – if you know what you're doing.

OUTSIDE EXERCISE

Whether you're jogging or walking, the dog at the end of the leash needs to have manners. You both should be able to enjoy a safe walk together, and that can only happen if you, the leader, are in control.

I don't believe in teaching a dog to heel. Heel is a throwback to World War II military training, when soldiers carried their rifles on their right side, while the dog (a four-legged soldier) had to stay on the left side, out of the rifle's line of fire. Unless you're a soldier, let's forego the militant heel and focus on how lovely it is to simply take a nice, peaceful walk with your best friend. But what do you do when your dog is walking you?

A client named Melissa showed me how she walked her dog, Tsarina, a beautiful white Siberian husky, a breed known for becoming sled dogs. The next thing I saw was Melissa's billowing orange skirt flapping behind her as Tsarina pulled her forcefully up the street, like a sled dog mushing her way through the snow. Melissa had to run faster and faster just to keep up with Tsarina. Clearly they needed help.

THE SUMO WRESTLER STOP

A dog pulls ahead for a simple reason – it works! He can speed you up a few extra feet or successfully drag you over to an interesting pile of poop that, according to him, really needs to be investigated. If he has slack in his leash, and you are not paying attention, he pulls, and as your hand extends, he gets the extra two feet he wants. Since Melissa was extending her hand when Tsarina pulled on her, she was escalating the problem by rewarding bad behavior. That would have to stop.

The first thing I did was take off Tsarina's choke chain. Melissa had been using it for months and had gotten no results, so why would we keep using it? Next, I put a leather belt around my waist and attached Tsarina's leash to it. Once the leash was attached to my belt, she could not extend it, no matter how hard she pulled. Talk about being connected! This arrangement also gave me both hands free to hold treats and toys.

Since the second issue was to get Tsarina to focus on me and not on everything but, I used her favorite toy, a squeaky tennis ball. I also pulled out one of my secret weapons – gold treats! I gave Tsarina a treat, and we started walking. Right off the bat, she walked directly beside me, focusing her attention on the bag of treats, which at this phase, I made sure she could see at all times. I also let her see the tennis ball, and she didn't pull on me even once. 'Walk, walk, good walk,' I praised her. She was learning that I had control of what she wanted – her favorite toy and treats. I never missed an opportunity to reward her with a treat when she came to check in with me, and

I gave the behavior a name. 'Close, close' I repeated in a singsong voice with an approving smile.

I wanted Tsarina to realize on her own that she had a certain length of leash that would remain the same. In the teaching phase, it's important *not* to use a retractable leash, so she gets familiar with her walking boundaries. Once a dog is good at walk and close, I can trust her with a longer leash, and maybe later on, with no leash at all. For now, though, when Tsarina pulled on her leash, she got an unexpected surprise. The walk stopped dead as I planted both of my feet on the ground beneath me like a Japanese sumo wrestler. I call this the sumo wrestler stop.

I did nothing and stood my ground, waiting for as long as it took Tsarina to figure out how to loosen the tension on her leash. After she took a step backward, Tsarina's leash was loose again, and we *immediately* started walking. If a dog is too strong for my sumo wrestler stop, I'll hold on to a nearby tree or a fence. I want the dog to understand the consequences of pulling. For the rest of the walk, Tsarina didn't try to pull me even once. But when the leash got hooked onto Melissa's belt, it was a different story.

I've seen dogs behave like angels around their trainers, but as soon as they get back with their owners again, it's all forgotten. I had no history with Tsarina, so our relationship was a blank slate. But since she had already established a pattern with Melissa, it took a bit longer. We needed to teach the dog a new pattern, and also eradicate the old association that her mom was a pushover.

We took two steps and Tsarina charged ahead, pulling on Melissa. Tsarina's attitude seemed to say, *It's only Mom at the end of the leash. I don't have to listen to her.* But Melissa was ready and assumed the sumo wrestler position, stopping as if she was glued to the ground. Then she waited. Melissa and Tsarina stood their ground for a moment or two. It was a standoff. I encouraged Melissa to act bored, as if she were saying, 'I've got all the time in the world, so go ahead and make up your mind. I'm not fighting with you anymore. It's up to you to loosen the leash.'

It can be a powerful tactic to act bored, but this 'talk to the hand' attitude is one of the easiest and best tools in The Loved Dog coaching arsenal. In a short period of time, Tsarina backed up and loosened the leash. Melissa had won and the walk continued as Melissa and I chanted, 'walk, walk'. We didn't offer Tsarina any treats, though, since we didn't want her to think that the pattern went: walk, charge ahead, stop, correct yourself, and get a treat. After a couple more tries, Tsarina was self-correcting very quickly, pulling less and less often, and getting treats for her correct behavior. Walking beside Melissa was now more fun than charging ahead!

YOUR DOG'S MESSAGE BOARD AND OTHER DISTRACTIONS

When you turn on your computer in the morning, don't you usually get a load of new and interesting information? I do, and I know that my dogs are looking for the same thing when we go on a walk. Your dog will want to catch up on all the news and gossip and check the 'p-mail' that other dogs have left for him, but he'll have to learn to wait patiently until you get to his 'message board'.

Even when Matt and Abby practiced the sumo wrestler stop, smart little Moby had devised a way to still get what he wanted. He would pull ahead when they were close to the curb or the bushes.

Although Matt stopped, Moby was in close enough range to check his p-mail, and he ended up getting rewarded for his unruly behavior.

I advised that when Moby charged ahead, Matt should take a large step to the side and then stop. Now Moby would not be able to get to his p-mail, and he would have to modify his bad manners, loosen the tension on the leash, and only then could he get to smell the latest gossip.

I use three different words on a walk:

1. **I say 'walk' when we are walking.**
2. **I say 'close' when the dog comes close to check in with me.**
3. **I say 'let's walk' when we are about to start or resume walking.**

I encouraged Matt and Abby to really focus on keeping Moby at their pace and not on his own. What if Abby was walking Moby and pushing the baby carriage at the same time? She might need to slow down for the safety of the baby, and Moby would have to alter his pace to match hers. To introduce this concept to Moby, I had Matt walk him on the leash and constantly change his pace. Moby would have to get in tune with his owner and pay attention to him, not to the voice in his head that wanted to go faster and faster.

It worked like a charm, and Moby did a little jump for joy when he understood what Matt and Abby wanted from him. It was as if he said, *I got it! I'm good.* Moby caught on fast, but some dogs take longer than others. Be patient, and it will work eventually, especially if you constantly praise your dog and talk to him while you're walking. Repeating the words 'good walk' or 'good close' can go a long way to encourage him. In the teaching phase, as always, give treats often. Of course, later on, you will switch to random rewards once your dog is a champion walker.

It helps to keep a positive, open frame of mind when you're teaching your dog this exercise. You need to be patient and avoid setting extra-high expectations. Your dog may not cover the distance you planned in a certain time. Try planning your walks a little bit differently. Instead of logging your success in miles or blocks, give yourself a time frame of fifteen or thirty minutes, and try to see how far you get. You may start out with one step and a sumo stop, and then manage to take two or three steps before you need to do the sumo stop once again. Recognize that your dog's success may need to start out small, but before you know it, you'll both be cruising the neighborhood together.

SUMMARY OF WALKING:

- **Leash length: keep the leash an ideal length of four to six feet.**
- **The sumo wrestler stop: when your dog charges ahead, stand firm without moving. Then wait patiently for your dog to give the leash some slack.**
- **Focus and praise: use the words 'walk', 'close', and 'let's walk' to encourage your dog along the way.**

Chapter Nineteen

Barking

I was at The Loved Dog Center, discussing a behavioral issue with my crew, when a passing siren caused a spontaneous karaoke session to break out among the dogs in our main room. They did not believe in one dog singing while the others listened, and every dog in the center joined the dissonant choir, each trying to be louder and more sincere than the dog next to him. It was like a pack of wolves howling at the moon, a bonding experience. But they weren't in Wyoming they were right in the middle of Los Angeles.

This singalong was not the first time I'd seen a group of dogs run a karaoke bar. When a chocolate Lab mix named Finn arrives at the center each morning, all the dogs greet him by barking out their joy. I call him the 'music teacher'. I often see one dog barking at another dog that he tries unsuccessfully to engage in a game. At times I hear a lonely dog howl for his owners, similar to the wild wolves howling in the desert.

Barking actually releases endorphins in a dog's system, in the same way that chocolate can affect people. Once the dog gets into it, he forgets why he even started barking. It just feels really good. Although it's an acceptable wolf behavior, excessive barking is bad manners for a domesticated dog, and if we don't help him break the habit, it's a surefire way to get tossed into a shelter. This kind of behavior is annoying to everyone and needs to be addressed since the dog is not only driving his owners crazy – he is driving the neighbors crazy too.

Scout, a West Highland terrier, loved to bark at anything that moved at any opportunity – not just when he didn't get what he wanted. He was so excitable and in love with barking that when I first arrived with my crew at his house to film him for my DVD, he would not stop crashing into pieces of equipment and making a royal pest of himself – and barking the entire time. He had quite the lung capacity and a vocabulary to match!

Since it was impossible for Scout's owners, Jane and Sara, to contain him and stop him from barking, I decided to start filming him as he rushed into the center of the room, uninvited. He stuck his head right into the camera lens, barking out his message to viewers. He was always underfoot, disrupting my crew and making so much noise that no one could hear himself think. Jane made a few feeble attempts to move Scout from the center of the chaos, but to no avail. She asked him quietly to stop barking, and she kept softly repeating his name so many times, I realized he had become desensitized to it. As far as Scout was concerned, Jane was boring, but my crew and their equipment were opportunities for exciting stimulation, which triggered his desire to bark.

He would not quiet down, even when Jane physically picked him up and moved him to a different place. The second she let him go, he was right back at it again, shouting to the rooftops. In order to help Scout fit into his owners' lives, I needed to teach him to associate barking with displeasure, and being quiet with pleasure. I decided that startling him was the right way to get him to stop. Then I could guide him to the proper behavior.

It was time for my trusty spray bottle.

While Scout was barking his head off, he suddenly felt a cool burst of water on the back of his head. He stopped barking instantly, shocked by this unexpected shower, and I quickly said, 'Shush. Good shush', and I fed him a gold treat. I also used a hand signal of a finger across my lips, the same signal I might use when I want a child to be quiet. He stayed quiet for a few more treats, and then, when a slithering wire got his attention, he barked again and chased it. Whoosh, the spray came again. This time he stopped and stared at me. I gave him a treat, repeating the words 'good shush', reinforcing that shushing was what I wanted him to do. He waited, considering what had just happened. I had increased his negative association with barking, and at the same time, I increased his pleasure association with shushing. Water is a harmless yet startling wake-up call that does no damage but instantly gets your dog's attention.

Later on that day, when Scout tried barking again, I simply used the hand signal to quiet him down, followed by praise for shushing and a few delicious treats. If he had not quietened down, I would have given him another short burst from the spray bottle. When I told Scout to shush and gave him the opportunity to comply before using the water bottle again, it turned out that I didn't have to use it at all. He was so smart, he got it with only two short sprays of water. Remember, the water was not a punishment, but negative reinforcement. I was prepared to keep bugging him with the water bottle as long as he kept bugging us with his barking. But it took only one spritz each time, and he shushed.

Now that Scout had gotten the point, I encouraged Jane and Sara to practice it with him throughout the day. When he was *not* barking, I asked them to approach him, use the hand signal, and tell him to shush, which he was already doing. Then they gave him a treat, one more way for Scout to associate shush with pleasure. Shush stopped him from barking the same way kisses stopped Honey from nipping.

For another approach (Plan B), try giving barking a name, like 'speak', and reward it with a bronze treat. You can ring the doorbell or tease your dog with a toy or a treat, and when he barks, say 'speak' and reward him. Once he learns to speak when you ask him to, then tell him to shush using the hand signal, and give him a gold treat for being quiet. He'll quickly realize that shushing pays off much better than speaking.

TELEPHONE BARKING

A client named Lillian called me to say that whenever she talked on the phone, her dog, a dachshund named Frida, would awaken from a dead sleep to start barking. This was a problem, especially because Lillian worked on the phone from home, selling health insurance.

'What do you do when Frida barks?' I asked her.

'I put the call on hold and yell at her to shut up,' Lillian said. 'Sometimes petting stops her, but I still have to get off the phone because she won't keep quiet. I can't do my job, and my dog is costing me money.'

There it was! Frida was being rewarded for bad behavior, so why should she stop? She'd been smart enough to figure out that when Lillian was on the phone and she barked, she could count on instant attention. Frida was no dummy. She had learned how to modify Lillian's behavior. But now it was Lillian's turn to modify Frida's behavior.

As usual, it would take practice, but Lillian did not want to practice when she was on an important work call. Just like getting fake guests to ring the doorbell to modify jumping, I suggested that Lillian should get her friends to make fake phone calls – or she could use her cell phone to call her house phone and pretend to be talking. I instructed her to ignore the barking and stay on the phone for a while. No yelling, scolding, or hanging up prematurely. If it got really bad, I suggested that Lillian leave the room and shut the door behind her, not looking at Frida and acting as if she weren't even there.

When she finally finished her call, I asked her to take Frida outside and play fetch, catch, or tug-of-war with her. A short doggy massage or a walk would also be good – anything that showered attention on Frida. Then when the phone rang again, Lillian was to ignore the dog completely. It was a short period of time before Frida began to understand that barking while her owner was on the phone got her absolutely nothing. It may require a lot of fake calls and ear plugs, just like teaching the dog not to jump might take a few fake guests. But it will work for you, just like it eventually did for Lillian. When Frida finally got it into her furry head that she was not in control and that Lillian *was*, the stage was set for good behavior.

SUMMARY OF BARKING:

- **Barking: teach your dog to shush with the help of a spray bottle and a hand signal. Reward him when he stops.**

- **Busy times: if your dog barks when you're busy or on the phone, don't reward his bad behavior or it will increase.**

Chapter Twenty

Celebrate!

My co-writer, Andrea, recently visited a woman who had just adopted a three-year-old black Labrador for her daughter, Carrie. But when Carrie started to show off what the 'trainer' had taught her, Andrea was devastated to see the prong collar go around the dog's neck. With her mother looking on, the little girl proceeded to take the dog into the backyard to try to get him to sit. The poor dog looked around as if he had never heard the word before, and then Carrie, a sweet little eight-year-old girl, yanked on the dog's neck like she'd been taught. The dog cowered and stumbled forward as Carrie gave his neck a harder yank, 'like the trainer told me to' she said with pride.

When the 'non-training' session was over and the dog was

released from his torture, he literally flew into Andrea's arms, looking at her as if to say, *Save me! Please!* After writing this book with me, Andrea understood that although they didn't realize it, Carrie and her mother were doing everything they could to turn a perfectly gentle, loving animal into a broken, aggressive, and potentially dangerous dog. And the poor thing still had no idea how to sit!

Although we have gained a great deal of awareness about being kind and respectful to animals, there are plenty of people who still abuse, choke, and beat up their animals and call it training. Please join me in becoming a dog advocate and spread the word that abuse is unnecessary and unacceptable. It was once considered rude and disruptive to interfere when someone was hitting a child in public. 'Turn the other way,' was the commonly accepted custom. 'It's none of your business.'

As a society, we have come to accept that abuse against children is *absolutely* our business, even if it's conducted in the name of 'discipline'. It should be the same with animals, too. We must all do whatever we can to end abusive dog training and to speak up for all living creatures who cannot speak up for themselves – whether they are homeless or living next door. Please interfere. Tell your friends, family, and acquaintances that they can choose a different way; a loving way to coach and play train their dogs. If we turn a blind eye to abuse, we are participating in it ourselves.

By now you and your dog have come a long way together! How far you go is up to you, and I hope you can see the wisdom in investing time in your dog and in your relationship together. Always remember that your dog is a diamond in the rough, waiting for you to unleash his magic. All you have to do is be patient, keep it fun, give him time to grasp the lessons, and reward every victory, large or small. We are so fortunate to have these wonderful creatures in our lives, and we have such a short time with them. Please respect them, appreciate them and celebrate them in each and every moment of your lives together!

Also available from Ebury Publishing

A Dog Abroad

One man and his dog journey into the heart of Europe

Bruce Fogle

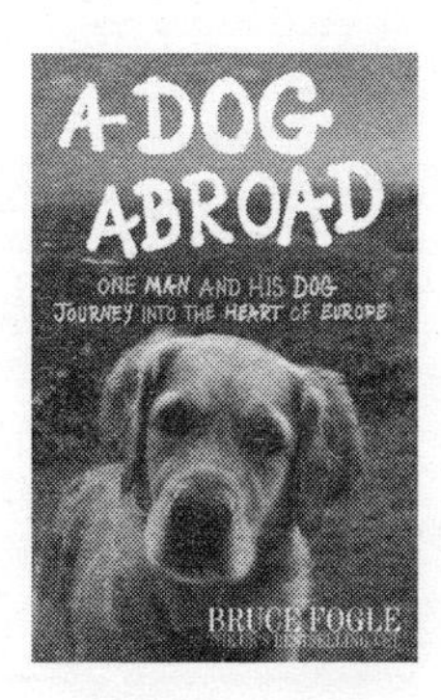

Travelling with man's best friend allows Bruce Fogle to savour landscape in a way he couldn't alone. Macy introduces him to an eclectic mix of rural folk and local animals, as well as other travellers. Together, Macy and Bruce walk for miles through peaceful woods and empty shores, taking time off from the frenetic bustle of urban life to appreciate the unspoilt beauty of landscapes still frozen in the aspic of time.

In this, their second adventure together, Bruce and Macy travel through the little known lands of Europe; tracking down the Fogle family roots. In their compact motorhome, they travel through the Netherlands, northern Germany and Scandinavia, the altics, and finally down the eastern rim of Poland to Slovakia and Slovenia. With Macy at his side, Fogle discovers new revelations about his ancestry as well as discovering the *real* central Europe.

The perfect travelling companion, *A Dog Abroad* is an enchanting read.

Also available from Ebury Publishing

Underdogs

Beauty is more than fur deep

Jim Dratfield

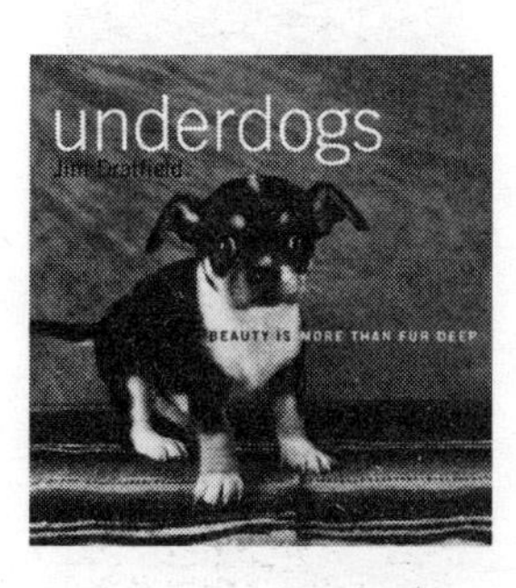

A charming, touching and occasionally hilarious celebration of man's best friend, in pictures and words, to cheer and inspire. If we love anything, it's the underdog - every kind of underdog. We recognise that beauty is more than fur deep. We appreciate dogs of indeterminate lineage, with mottled fur, too-large ears, or strange little bodies; we love dogs who look like impish little children or grumpy old men.

Underdogs is their book - the character displayed in the funny faces and singular shapes, uncommon breeds and unrecognised mixed-breeds is a reflection of the true companionship we really love about dogs.

Also available from Ebury Publishing

Dogs: Homoeopathic Remedies

George Macleod

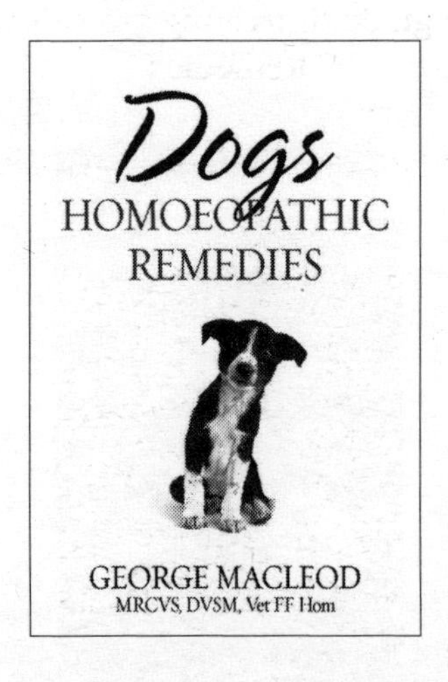

This comprehensive guide introduces the principles of homoeopathy and the nature of homoeopathic remedies, explaining how remedies can be prepared and administered. There are informative sections on treating the different canine bodily systems and the specific diseases that dogs may suffer from. There is also helpful advice on canine virus and bacterial diseases, as well as the diseases of puppyhood, making this book a must for any dog owner.

Also available from Ebury Publishing

Bach Flower Remedies for Animals

The definitive guide to treating animals with the Bach Remedies

Stefan Ball and Judy Ramsell Howard

Bach Flower Remedies for Animals is a complete and authoritative guide to using the Bach Flower Remedies to treat pets and other animals. Written by experts, it includes the history behind Dr Bach's internationally acclaimed remedies and explains how the remedies are suitable for animals. According to an animal's temperament, a particular remedy or combination of remedies will prove the most beneficial for it.

There is guidance on using the remedies and on reading animal behaviour, drawing on insights from professional animal behaviourists. There is also detailed information on treating animals in the home, horses, and animals on the farm or in the wild. Packed full of helpful advice, there are intriguing case studies throughout.